AF595007

UAS-Remote Sensing Methods for Mapping, Monitoring and Modeling Crops

UAS-Remote Sensing Methods for Mapping, Monitoring and Modeling Crops

Editors

Francisco Javier Mesas Carrascosa
Joaquim João Sousa

MDPI • Basel • Beijing • Wuhan • Barcelona • Belgrade • Manchester • Tokyo • Cluj • Tianjin

Editors
Francisco Javier Mesas Carrascosa
University of Cordoba
Spain

Joaquim João Sousa
Universidade de Trás-os-Montes e Alto Douro
Portugal

Editorial Office
MDPI
St. Alban-Anlage 66 4052 Basel,
Switzerland

This is a reprint of articles from the Special Issue published online in the open access journal *Remote Sensing* (ISSN 2072-4292) (available at: https://www.mdpi.com/journal/remotesensing/special_issues/UAS_remote_sensing_methods_mapping_monitoring_modeling_crops).

For citation purposes, cite each article independently as indicated on the article page online and as indicated below:

LastName, A.A.; LastName, B.B.; LastName, C.C. Article Title. *Journal Name* **Year**, *Volume Number*, Page Range.

ISBN 978-3-0365-0526-8 (Hbk)
ISBN 978-3-0365-0527-5 (PDF)

Contents

About the Editors

Francisco Javier Mesas Carrascosa (Dr. in Engineering, Engineer in Geodesy and Cartography) is Senior Lecturer at the Higher Technical School of Agricultural Engineering of University of Córdoba, Spain. His scientific research has been centered on the use of geomatics techniques for precision agriculture, mainly on the use of remotely sensed imagery acquired by sensors aboard satellite, manned, and unmanned platforms.

Joaquim João Sousa (Ph.D. in Surveying Engineering) is Senior Lecturer at the School of Sciences and Technology of University of Trás-os-Montes e Alto Douro. His main research interests focus on the use of unmanned aerial vehicle (UAV) imagery (RGB, NIR, thermal and multi- and hyperspectral) for agriculture and forestry applications.

Editorial

UAS-Remote Sensing Methods for Mapping, Monitoring and Modeling Crops

Francisco Javier Mesas-Carrascosa

Department of Graphic Engineering and Geomatics, Campus de Rabanales, University of Cordoba, 14071 Cordoba, Spain; fjmesas@uco.es

Received: 12 November 2020; Accepted: 24 November 2020; Published: 26 November 2020

Abstract: The advances in Unmanned Aerial Vehicle (UAV) platforms and on-board sensors in the past few years have greatly increased our ability to monitor and map crops. The ability to register images at ultra-high spatial resolution at any moment has made remote sensing techniques increasingly useful in crop management. These technologies have revolutionized the way in which remote sensing is applied in precision agriculture, allowing for decision-making in a matter of days instead of weeks. However, it is still necessary to continue research to improve and maximize the potential of UAV remote sensing in agriculture. This Special Issue of Remote Sensing includes different applications of UAV remote sensing for crop management, covering RGB, multispectral, hyperspectral and LIght Detection and Ranging (LiDAR) sensor applications on-board (UAVs). The papers reveal innovative techniques involving image analysis and cloud points. It should, however, be emphasized that this Special Issue is a small sample of UAV applications in agriculture and that there is much more to investigate.

Keywords: RGB; multispectral; hyperspectral; LiDAR; agriculture

---

1. Introduction

In order to satisfy the needs of the global population while considering the reduction of agricultural areas, investments in agri-food sectors have grown with the goal of increasing productivity by at least 70% by 2050 [1]. Emerging technologies such as Internet of the Things as well as new methods for analyzing data to reveal patterns and trends improve the potential of Precision Agriculture (PA) and enable the improvement of productivity. Among these technologies, Remote Sensing (RS) is considered to be one of the most important for this purpose. In the last four decades, RS has been used to monitor crops [2], using, initially, images from sensors on-board satellite platforms. The spatial, temporal and spectral resolutions required for many PA applications are limited when using these platforms, mainly in woody crops. RS sensors on-board UAVs have provided an interesting alternative to fill this gap, meeting the ultra-high requirements of these resolutions.

The purpose of this Special Issue is to promote the new developments in Unmanned Aerial System–Remote Sensing (UAS-RS) methods in PA for the mapping, monitoring and modeling of crops.

2. Overview of Contributions

This Special Issue includes advances based on RGB [3–5] multispectral [6], hyperspectral [7], LiDAR [8] and hyperspectral-LiDAR [9] sensors on-board UAVs, focusing on applications related to PA. These approaches applied various methodologies based on both images and cloud points.

Jurado et al. [3] present a method to detect and locate individual grapevine trunks using 3D cloud points from UAV-based RGB images. Their major contribution is a fully automatic approach, which does not require any prior knowledge of the number of plants per row. In addition, the computational complexity does not demand high-performance computing. Moreover, they conclude that their approach can be extended to estimate other biophysical parameters of grapevines, with the final goal being to provide efficient vineyard management.

Ronchetti et al. [4] use RGB and multispectral images from UAV flights for crop row detection. They tested and compared different methodologies based on Bayesian segmentation, thresholding algorithms and classification algorithms, which were applied on vineyards, pear orchards and tomato fields. Although Digital Surface Model (DSM), RGB and multispectral ortomosaics offered adequate results, for crops characterized by high heights, like vineyards and pear crops, DSM as input offered better results. Therefore, RGB sensors on-board UAVs can be an alternative to a more expensive sensor, like multispectral.

Feng et al. [5] propose an Attention-based Recurrent Convolutional Neural Network (ARCNN) for mapping crops from multi-temporal RGB-UAV images to obtain useful phenological information. Overall accuracy and Kappa coefficient in classification processing were high despite the low spectral resolution of the sensor used. This model could be understood as a general framework for multi-temporal RS image processing.

Using multispectral sensors on-board a UAV, Jełowicki et al. [6] estimate losses in rapeseed crops. Three vegetation indexes were evaluated, the Normalized Difference Vegetation Index (NDVI), Soil-Adjusted Vegetation Index (SAVI) and Optimized Soil-Adjusted Vegetation Index (OSAVI), calculated using red and near infrared spectral regions, and red edge and near infrared. They conclude that a ground sample distance equal to 10 cm is enough to detect damaged areas, which means higher flight altitude and therefore increased area covered. Regarding vegetation indexes, the OSAVI calculated using red edge and near infrared offered better results to monitor crop condition.

While multispectral sensors are very useful for crop monitoring, there are occasions in which it is necessary to use narrow spectral bandwidth, using hyperspectral sensors with high spectral resolution. Santos et al. [7] evaluate different wavelength selection methods based on the partial least squares (PLS) method. The objective was to select the best wavelength to classify two irrigation systems used in olive orchards. The variation in the evaluated methods showed the need to select the appropriate method in a case by case scenario. In their study, the Genetic Algorithm PLS, Regression Coefficient PLS, Forward Interval PLS, Lasso, Boruta and All-together methods showed the most promising results, offering an overall accuracy of 75% or higher in the classification.

In addition to passive sensors, active LiDAR sensors allow the generation of dense 3D point clouds. Today, with the miniaturization of the sensors and the reduction in weight, it is possible to apply these systems to UAV platforms. The correct three-dimensional modelling of a crop requires a dense and accurate point cloud. Chen et al. [8] present a methodology for an integrated navigation and positioning optimization method based on the grasshopper optimization algorithm and a point cloud density enhancement method.

Finally, although the use of singular sensors on-board UAVs offers very interesting data to be used in PA, it is even more interesting if data from different sensors are combined. Masjedi et al. [9] explore the potential for reliable prediction of sorghum biomass using multi-temporal hyperspectral and LiDAR data acquired by sensors mounted on UAV platforms. Among all the derived variables, nitrogen- and photosynthesis-related features extracted from hyperspectral data and geometric based features derived from the LiDAR data were the most interesting. In addition, they evaluated the most appropriate date for data collection after the sowing in order to improve the results of the predictive models.

Applications for UAS in agriculture have progressed significantly in recent years as the technology has improved in tandem with decreasing costs. Motivating a dual interest from farmers and businesses in UAS, the interest for these technologies has grown in applied agriculture, with new applications being developed in the agri-food sector. Most applications rely on the ability to generate and deliver precise and accurate information to support agricultural activities or to inform complementary activities like crop analysis and monitoring. Consequently, data quality is important and is the core priority of drone use decisions. Given the relative infancy of agricultural UAS technology, there is still much progress and research to be made. Despite this, it is really only a matter of time until this UAS technology is mature enough to act as a replacement for existing methods as the industry is rapidly integrating newer sensors and processing technologies, constantly improving the quality of the data captured.

Funding: This research received no external funding.

Conflicts of Interest: The author declares no conflict of interest.

References

1. FAO. *Declaration of the World Summit on Food Security*; Food and Agriculture Organization of the United Nations: Rome, Italy, 2009.
2. Mulla, D.J. Twenty five years of remote sensing in precision agriculture: Key advances and remaining knowledge gaps. *Biosyst. Eng.* **2013**, *114*, 358–371. [CrossRef]
3. Jurado, J.M.; Pádua, L.; Feito, F.R.; Sousa, J.J. Automatic Grapevine Trunk Detection on UAV-Based Point Cloud. *Remote Sens.* **2020**, *12*, 3043. [CrossRef]
4. Ronchetti, G.; Mayer, A.; Facchi, A.; Ortuani, B.; Sona, G. Crop Row Detection through UAV Surveys to Optimize On-farm Irrigation Management. *Remote Sens.* **2020**, *12*, 1967. [CrossRef]
5. Feng, Q.; Yang, J.; Liu, Y.; Ou, C.; Zhu, D.; Niu, B.; Liu, J.; Li, B. Multi-Temporal Unmanned Aerial Vehicle Remote Sensing for Vegetable Mapping Using an Attention-Based Recurrent Convolutional Neural Network. *Remote Sens.* **2020**, *12*, 1668. [CrossRef]
6. Jełowicki, Ł.; Sosnowicz, K.; Ostrowski, W.; Osińska-Skotak, K.; Bakuła, K. Evaluation of Rapeseed Winter Crop Damage Using UAV-Based Multispectral Imagery. *Remote Sens.* **2020**, *12*, 2618. [CrossRef]
7. Santos-Rufo, A.; Mesas-Carrascosa, F.-J.; García-Ferrer, A.; Meroño-Larriva, J.E. Wavelength Selection Method Based on Partial Least Square from Hyperspectral Unmanned Aerial Vehicle Orthomosaic of Irrigated Olive Orchards. *Remote Sens.* **2020**, *12*, 3426. [CrossRef]
8. Chen, J.; Zhang, Z.; Zhang, K.; Wang, S.; Han, Y. UAV-Borne LiDAR Crop Point Cloud Enhancement Using Grasshopper Optimization and Point Cloud Up-Sampling Network. *Remote Sens.* **2020**, *12*, 3208. [CrossRef]
9. Masjedi, A.; Crawford, M.M.; Carpenter, N.R.; Tuinstra, M.R. Multi-Temporal Predictive Modelling of Sorghum Biomass Using UAV-Based Hyperspectral and LiDAR Data. *Remote Sens.* **2020**, *12*, 3587. [CrossRef]

Article

Multi-Temporal Unmanned Aerial Vehicle Remote Sensing for Vegetable Mapping Using an Attention-Based Recurrent Convolutional Neural Network

Quanlong Feng [1,2], Jianyu Yang [2,*], Yiming Liu [2], Cong Ou [2], Dehai Zhu [2], Bowen Niu [2], Jiantao Liu [3] and Baoguo Li [2]

1 College of Resources and Environmental Sciences, China Agricultural University, Beijing 100193, China; fengql@cau.edu.cn

2 College of Land Science and Technology, China Agricultural University, Beijing 100083, China; liuym0086@cau.edu.cn (Y.L.); oucong@cau.edu.cn (C.O.); zhudehai@cau.edu.cn (D.Z.); s20193081417@cau.edu.cn (B.N.); libg@cau.edu.cn (B.L.)

3 School of Surveying and Geo-Informatics, Shandong Jianzhu University, Jinan 250101, China; liujiantao18@sdjzu.edu.cn

* Correspondence: ycjyyang@cau.edu.cn; Tel.: +86-10-62737554

Received: 19 April 2020; Accepted: 21 May 2020; Published: 22 May 2020

Abstract: Vegetable mapping from remote sensing imagery is important for precision agricultural activities such as automated pesticide spraying. Multi-temporal unmanned aerial vehicle (UAV) data has the merits of both very high spatial resolution and useful phenological information, which shows great potential for accurate vegetable classification, especially under complex and fragmented agricultural landscapes. In this study, an attention-based recurrent convolutional neural network (ARCNN) has been proposed for accurate vegetable mapping from multi-temporal UAV red-green-blue (RGB) imagery. The proposed model firstly utilizes a multi-scale deformable CNN to learn and extract rich spatial features from UAV data. Afterwards, the extracted features are fed into an attention-based recurrent neural network (RNN), from which the sequential dependency between multi-temporal features could be established. Finally, the aggregated spatial-temporal features are used to predict the vegetable category. Experimental results show that the proposed ARCNN yields a high performance with an overall accuracy of 92.80%. When compared with mono-temporal classification, the incorporation of multi-temporal UAV imagery could significantly boost the accuracy by 24.49% on average, which justifies the hypothesis that the low spectral resolution of RGB imagery could be compensated by the inclusion of multi-temporal observations. In addition, the attention-based RNN in this study outperforms other feature fusion methods such as feature-stacking. The deformable convolution operation also yields higher classification accuracy than that of a standard convolution unit. Results demonstrate that the ARCNN could provide an effective way for extracting and aggregating discriminative spatial-temporal features for vegetable mapping from multi-temporal UAV RGB imagery.

Keywords: vegetable mapping; multi-temporal UAV; recurrent convolutional neural network; attention mechanism

---

1. Introduction

Accurate vegetable mapping is of great significance for modern precision agriculture. The spatial distribution map for different kinds of vegetables is the basis for automated agricultural activities such as unmanned aerial vehicle (UAV)-based fertilizer and pesticide spraying. Traditional vegetable mapping is usually based on field survey or visual interpretation of remote sensing imagery, which is time-consuming and inconvenient. Hence, it is of great importance to study the automatic methods for precise vegetable classification. However, previous studies mainly focused on the staple crop (e.g., corn, paddy rice) classification [1], in this regard, we are highly motivated to propose an effective method for vegetable classification based on UAV observations, which could provide a useful reference for future studies on vegetable mapping.

In previous studies, optical satellite imagery was firstly utilized for vegetable and crop mapping. Wikantika et al. applied linear spectral mixture analysis to map vegetable parcels from mountainous regions based on Landsat Enhanced Thematic Mapper (ETM) data [2]. Belgiu et al. utilized multi-temporal Sentinel-2 imagery for crop mapping based on a time-weighted dynamic time warping method and achieved comparable accuracy with random forest (RF) [3]. Rupasinghe et al. adopted Pleiades data and a support vector machine (SVM) to classify the coastal vegetation cover and also yielded a good classification performance [4]. Wan et al. also used SVM and single-date WorldView-2 for crop type classification and justified the role of texture features in improving the classification accuracy [5]. Meanwhile, as the new generation sensor of Landsat satellite, data acquired by the Operational Land Imager (OLI) from Landsat-8 has also been used for vegetable and crop type classification. For instance, Asgarian et al. used multi-date OLI imagery for vegetable and crop mapping in central Iran based on decision tree and SVM and achieved good results [6].

However, when compared with staple crops, the land parcel of vegetables is small in size, resulting in a large amount of mixed pixels in space-borne images. Different from space-borne observations, a UAV could obtain images with very high or ultra-high spatial resolution where the mixed pixel is no longer a problem. Meanwhile, a UAV could be deployed whenever necessary, which makes it an efficient tool for rapid monitoring of land resources [7–11]. Due to payload capacity limitations, off-the-shelf digital cameras have been equipped in small-sized UAVs [7,8,11]. Under this circumstance, the images acquired only have three bands (i.e., red, green, blue, RGB), resulting in a low spectral resolution which limits the performance of differentiating various vegetable categories. To reduce this impact, we introduce multi-temporal UAV data to obtain useful phenological information to enhance the inter-class separability. Afterwards, a robust classification model, the attention-based recurrent convolutional neural network (ARCNN), is constructed to further improve the classification accuracy.

Compared with mono-temporal or single-date observation, multi-temporal datasets could provide useful phenological information, which aids for plant and vegetation classification during growing season [10–15]. Pádua et al. adopted multi-temporal UAV-based RGB imagery to differentiate grapevine vegetation from other plants in a vineyard [11], and indicates that although RGB images have a low spectral resolution, the inclusion of multi-temporal observations makes it possible for accurate plant classification. Van Iersel conducted river floodplain vegetation classification using multi-temporal UAV data and a hybrid method, which is based on the combination of random forest and object-based image analysis (OBIA) [14]. In our previous research, we also incorporated multi-temporal Landsat imagery and a series of classifiers (i.e., decision trees, SVM and RF) for cropland mapping of the Yellow River delta [15], which justified the effectiveness of multi-temporal observations in enhancing the classification performance.

The above mentioned studies are mainly based on low-level, manually designed features (i.e., spectral indices, textures) and machine-learning classifiers, which might show poor performance in obtaining the high-level and representative features for accurate vegetable mapping. Besides, lots of domain expertise together with engineering skills are always involved in these manually designed features [16,17]. Meanwhile, deep learning offers a novel way for discovering informative features through a hierarchical learning framework [18], which shows promising performance in several computer vision (CV) applications such as sematic segmentation [19,20], object detection [21] and image classification [22–24]. Recently, deep learning models have been widely studied in the field of remote sensing [25–29], including cloud detection [30], building extraction [31–33], land object detection [34,35], scene classification [36,37] and land cover mapping [38–43]. Specifically, Kussul et al.

utilized both one- and two-dimensional CNN to classify crop and land cover types from both synthetic aperture radar (SAR) and optical satellite imagery [41]. Rezaee et al. [42] also adopted an AlexNet [22] pre-trained on ImageNet for wetland classification based on mono-temporal RapidEye optical data. It should be noted that the vegetable land parcels in China are always mixed with other crops and the landscape is rather fragmented, leading to a large variability in the shape and scale of land parcels [44]. However, previous studies usually mirror deep learning models from computer vision field while neglecting the complex nature of agricultural landscapes. Therefore, how to build an effective deep learning model to account for the fragmented landscape is a key issue in vegetable mapping.

Meanwhile, the introduction of multi-temporal UAV data calls for effective methods of temporal feature extraction and spatial-temporal fusion to further boost the classification accuracy. Early studies [11,12,14] usually stacked or concatenated multi-temporal data without considering the hidden temporal dependencies. With the development of recurrent neural network (RNN), models such as long-short term memory (LSTM) [45] have been adopted to establish the relationship between sequential remote sensing data [46–50]. Ndikumana et al. utilized multi-temporal SAR Sentinel-1 imagery and a RNN for crop type classification, and indicated that the RNN showed a higher accuracy than several popular machine learning models (i.e., RF and SVM) [46]. Mou et al. cascaded a CNN and RNN to aggregate spectral, spatial and temporal features for change detection [47]. In addition, when it comes to vegetable or crop mapping, it should be noted that the importance or contribution of each mono-temporal dataset to classification may vary during the growing season. Therefore, how to model the sequential relationship between multi-temporal UAV data hence to further boost the classification accuracy is of great significance.

To tackle the above issues, this study proposes an attention-based recurrent convolutional neural network (ARCNN) for accurate vegetable mapping from multi-temporal UAV data. The proposed model integrates a multi-scale deformable CNN and an attention-based RNN into a trainable end-to-end network. The former is to learn and extract the representative spatial features from UAV data to account for the scale and shape variations under fragmented agricultural landscape, while the latter is to model the dependency across multi-temporal images to obtain useful phenological information. The proposed network yields an effective solution for spatial-temporal feature fusion, based on which the vegetable mapping accuracy could be boosted.

2. Materials and Methods

2.1. Study Area

Both the study area and multi-temporal UAV imagery used in this research are illustrated in Figure 1.

The study area includes a vegetable field which is located in Xijingmeng Village of Shenzhou City, Hebei province, China. There are various kinds of vegetables, such as Chinese cabbage, carrot, leaf mustard, etcetera. Meanwhile, the study area also locates in the North China Plain, which belongs to a continental monsoon climate, where summer is humid and hot, while winter is dry and cold. The annual temperature is about 13.4 °C and the annual precipitation is about 486 mm. Vegetables are usually planted in late August and harvested in early November.

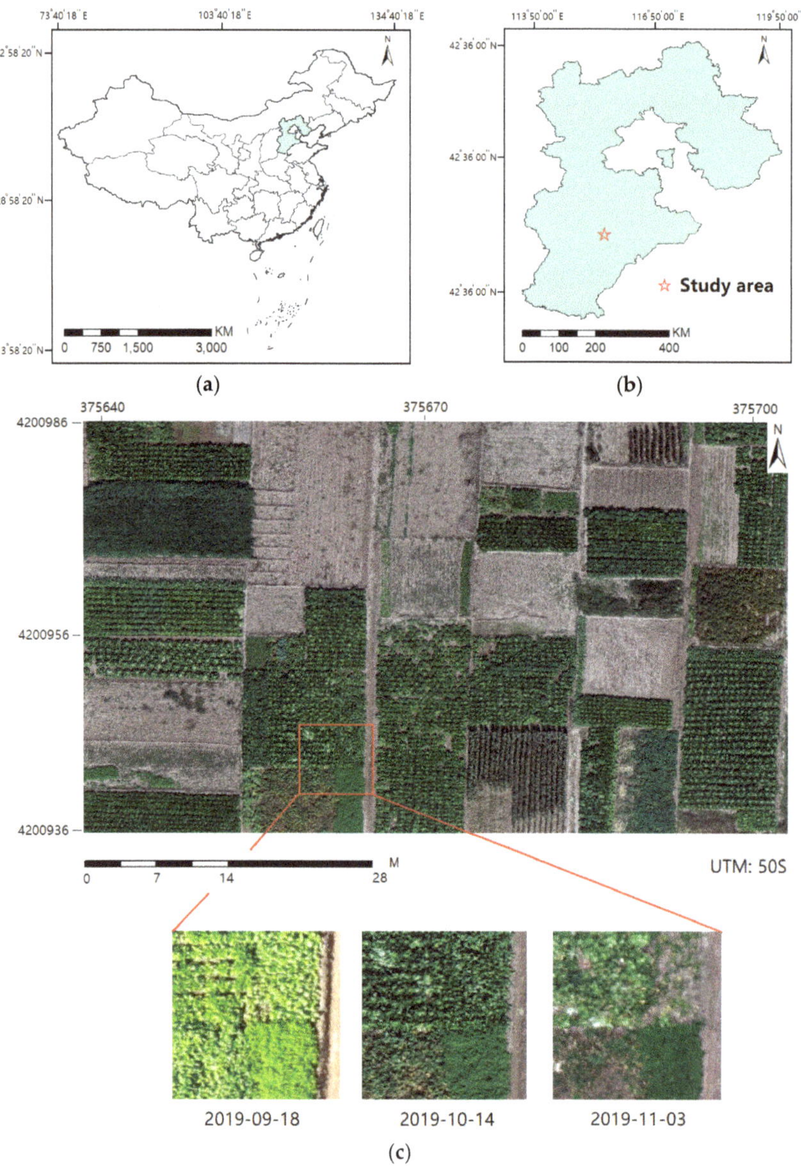

Figure 1. Study area. (**a**) China; (**b**) Hebei Province; and (**c**) Xijingmeng Village and multi-temporal UAV images.

A field survey was conducted along with the UAV flight. Vegetable and crop types, locations measured by global positioning system (GPS) and photographs were recorded for every land parcel. According to the results of the field survey, there were a total of fourteen land cover categories, including eight vegetable types (i.e., carrot, Chinese cabbage, leaf mustard, turnip, spinach, kohlrabi, potherb and scallion), four crop types (i.e., millet, sweet potato, corn and soybean), weed and bare soil (Table 1).

Table 1. Classification scheme of this study.

No.	Class Name	Training/Testing	Ground Image	No.	Class Name	Training/Testing	Ground Image
1	Carrot	200/200		8	Millet	200/200	
2	Chinese cabbage	400/400		9	Weed	100/100	
3	Leaf mustard	200/200		10	Bare soil	200/200	
4	Turnip	200/200		11	Sweet potato	200/200	
5	Spinach	50/50		12	Corn	50/50	
6	Kohlrabi	50/50		13	Soybean	200/200	
7	Potherb	100/100		14	Scallion	100/100	

Training and testing datasets were obtained from UAV imagery by visual inspection based on the sampling sites' GPS coordinates and the corresponding land cover categories. Numbers of both training and testing datasets are shown in Table 1. Besides, Table 1 also shows the ground image taken during the field work to depict the detailed appearance of various vegetables and crops.

Meanwhile, Figure 2 illustrates the spatial distribution of both training and testing samples. It indicates that all the samples are randomly distributed and no overlap exists between training and testing regions. Besides, because we adopted patch-based per-pixel classification, all the training and testing samples are pixels from the region of interest (ROI). In this study, the number of training and testing samples are both 2250, respectively, which accounts for a small area (0.03%) of the total study region (7,105,350 pixels).

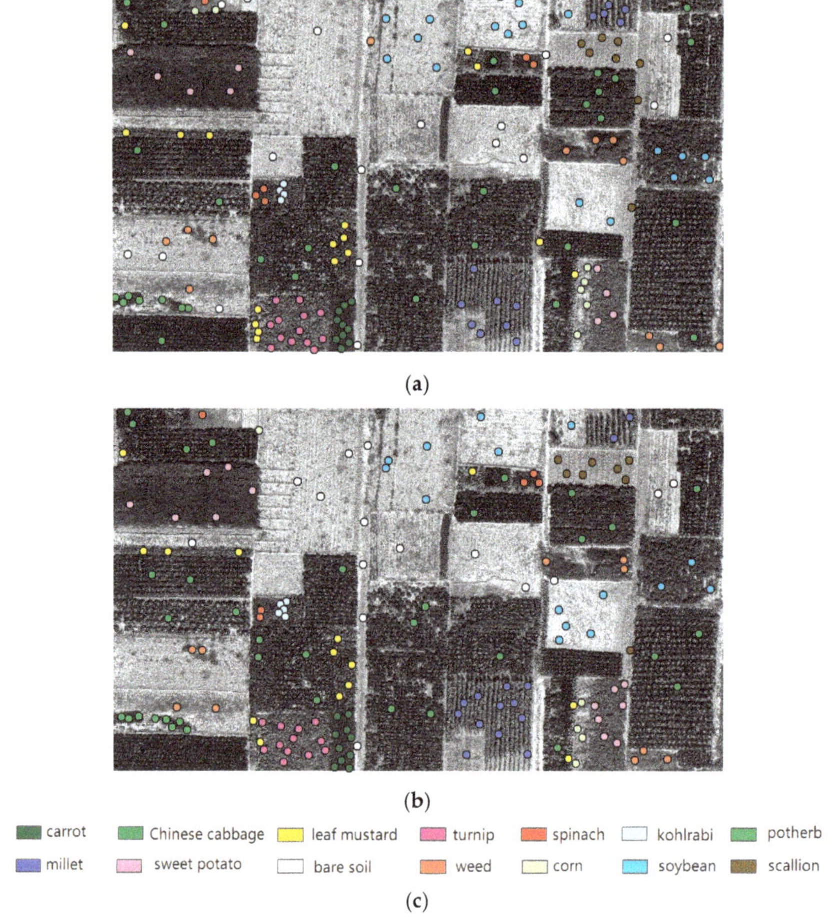

Figure 2. Spatial distribution of (**a**) training samples; (**b**) testing samples; (**c**) legend.

2.2. Dataset Used

We utilized a small-sized UAV, DJI-Inspire 2 [51], for the image data acquisition. The camera onboard is an off-the-shelf, light-weight digital camera with only three RGB bands. Therefore, the low spectral resolution would make it difficult to separate various vegetable categories if only considering single-date UAV data. To tackle this issue, we introduce multi-temporal UAV observations, which could obtain the phenological information during the growing season to increase the inter-class separability.

We conducted three flights in the autumn of 2019 (Table 2). During each flight, the flying height was set to be 80 m, achieving a very high spatial resolution of 2.5 cm/pixel. Besides, the width and height of the study area is 3535 and 2010 pixels (88.4 m and 50.3 m), respectively. Actually, the extent of the study area is at the limit of UAV data coverage. Although the study area may still seem small, it is limited by the operation range of the mini-UAV used. In future study, we would try high altitude long endurance (HALE) UAV to acquire images of a larger study region.

Table 2. Multi-temporal UAV images utilized in this study.

	Season	Date	Data Source
T1	Autumn	18 September 2019	UAV RGB data
T2	Autumn	14 October 2019	UAV RGB data
T3	Autumn	3 November 2019	UAV RGB data

The raw images acquired during each flight were orthorectified firstly and then mosaicked to an entire image by Pix4D [52]. Specifically, several key parameters in Pix4D are set as follows. "Aerial Grid or Corridor" is chosen for matching image pairs, "Automatic" is selected for targeted number of key points, matching window size is 7 × 7 and 1 GSD is used for resolution. The rest of the parameters are set to default values. Afterwards, image registration was performed among the multi-temporal UAV data by ENVI (the Environment for Visualizing Images) [53].

2.3. Overview of the ARCNN

Figure 3 illustrates the architecture of the proposed attention-based recurrent convolutional neural network (ARCNN) for vegetable mapping from multi-temporal UAV data. It mainly contains two parts, (1) a spatial feature extraction module based on a multi-scale deformable convolutional network (MDCN), and (2) a spatial-temporal feature fusion module based on a bi-directional RNN and attention mechanism. The former is to learn representative spatial features while the latter is to aggregate spatial and temporal features for the final vegetable classification.

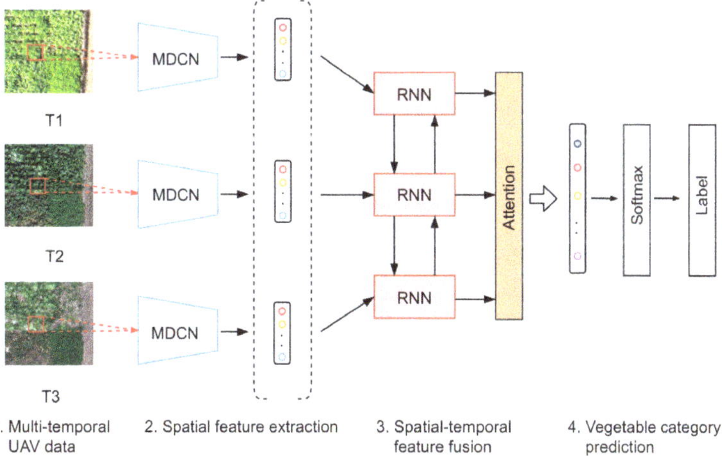

Figure 3. The overview of the proposed attention-based recurrent convolutional neural network (ARCNN) for vegetable mapping based on multi-temporal UAV data.

2.4. Spatial Feature Extraction Based on MDCN

Accurate vegetable classification requires discriminative features. In this section, a multi-scale deformable convolutional network (MDCN) is proposed to learn and extract rich spatial features from UAV imagery, which is to account for the scale and shape variations of land parcels. Specifically, MDCN is an improved version of our previous study [44], and the network structure is depicted as Figure 4.

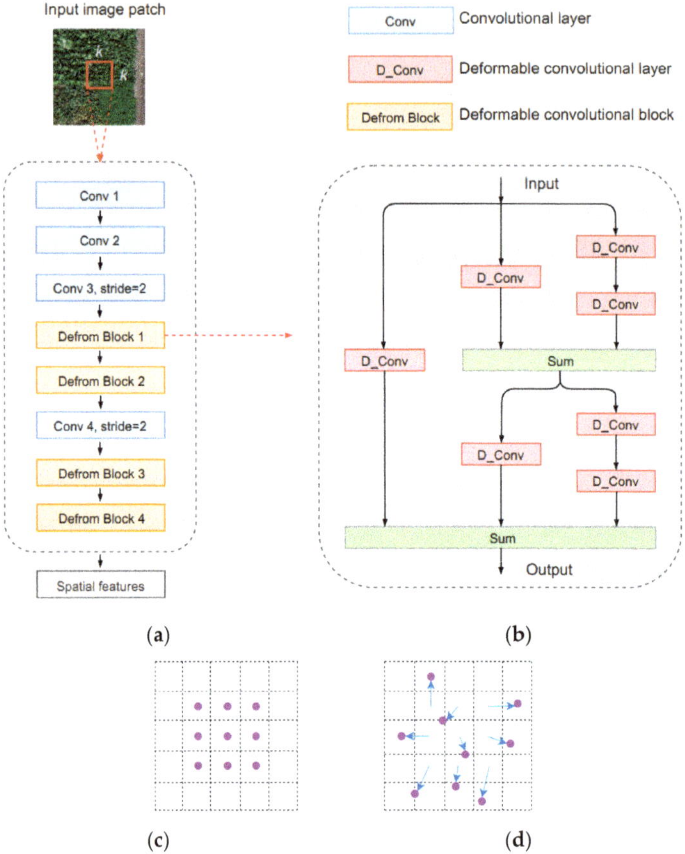

Figure 4. (**a**) Network structure of the proposed multi-scale deformable convolutional network (MDCN); (**b**) deformable convolutional block; (**c**) standard convolution; and (**d**) deformable convolution.

Same as our previous work, the input of MDCN is an image patch which is located at the center of the labeled pixel. The dimension of the patch is $k \times k \times c$ [44], where k stands for the patch size while c refers to the channel number. Specifically, MDCN includes four regular convolutional layers and four deformable convolutional blocks. Table 3 shows the detailed configuration of the MDCN.

Table 3. Detailed configuration of the MDCN.

Layer Name	Input Size	Output Size	Kernel Size	Filter Number	Stride
Input	$11 \times 11 \times 3$	–	–	–	–
Conv1	$11 \times 11 \times 3$	$11 \times 11 \times 64$	3×3	64	1
Conv2	$11 \times 11 \times 64$	$11 \times 11 \times 128$	3×3	128	1
Conv3	$11 \times 11 \times 128$	$6 \times 6 \times 128$	3×3	128	2
Deform Block 1	$6 \times 6 \times 128$	$6 \times 6 \times 128$	–	–	–
Deform Block 2	$6 \times 6 \times 128$	$6 \times 6 \times 128$	–	–	–
Conv4	$6 \times 6 \times 128$	$3 \times 3 \times 128$	3×3	256	2
Deform Block 3	$3 \times 3 \times 256$	$3 \times 3 \times 256$	–	–	–
Deform Block 4	$3 \times 3 \times 256$	$3 \times 3 \times 256$	–	–	–

The deformable block contains multiple streams of deformable convolution [54], which could learn hierarchical and multi-scale features. The role of deformable convolution is to model the shape variations under complex agricultural landscapes. Considering that the standard convolution only samples the given feature map at fixed locations [54,55], it could not handle the geometric transformations. Compared with standard convolution, deformable convolution introduces additional offsets along with the standard sampling grid [54,55], which could account for various transformations for scale, aspect ratio and rotation, making it an ideal tool to extract robust features under complex landscapes. During the training process, both the kernel and offsets of a deformable convolution unit can be learned without additional supervision. In this situation, the output y at the location p_0 could be calculated according to Equation (1):

$$y(p_0) = \sum w(p_i) * x(p_0 + p_i + \Delta p_i) \tag{1}$$

where w stands for the learned weights, p_i means the i th location, x represents the input feature map and Δp_i refers to the offset to be learned [54]. In addition, as for the determination of the patch size k, we referred to our previous research [44] and the highest classification performance was reached when k equaled 11.

2.5. Spatial-Temporal Feature Fusion

After the extraction of spatial features from every mono-temporal UAV image, it is essential to establish the relationship between these sequential features to yield a complete feature representation for boosting the vegetable classification performance. In this section, we exploit an attention based bi-directional LSTM (Bi-LSTM-Attention) for the fusion of spatial and temporal features (Figure 5). The network structure of Bi-LSTM-Attention is illustrated as follows.

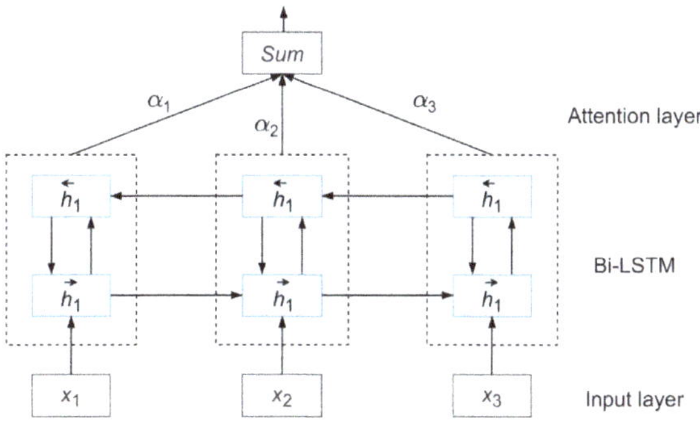

Figure 5. Architecture of the attention-based bi-directional LSTM.

Specifically, LSTM is a variant of RNN, which contains one input layer, one or several hidden layers and one output layer [45]. It should be noted that LSTM is more specialized in capturing long-range dependencies between sequential signals than other RNN models. LSTM utilizes a vector (i.e., memory cell) to store the long-term memory and adopts a series of gates to control the information flow [45] (Figure 6). The hidden layer is updated as follows:

$$i_t = \sigma(W_{ix}x_t + W_{ih}h_{t-1} + b_i) \tag{2}$$

$$f_t = \sigma(W_{fx}x_t + W_{fh}h_{t-1} + b_f) \tag{3}$$

$$c_t = f_t c_{t-1} + i_t \tanh(W_{cx}x_t + W_{ch}h_{t-1} + b_c) \tag{4}$$

$$o_t = \sigma(W_{ox}x_t + W_{oh}h_{t-1} + b_o) \tag{5}$$

$$h_t = o_t \tanh(c_t) \tag{6}$$

where i refers to the input gate, f stands for the forget gate, o refers to the output gate, c is the memory cell and σ stands for the logistic sigmoid function [45].

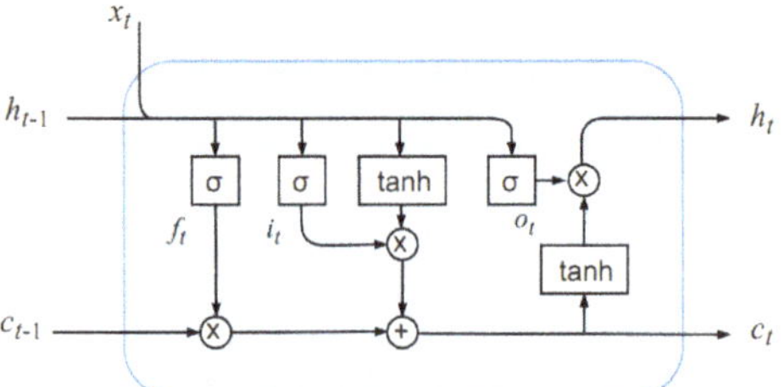

Figure 6. Structure of the LSTM.

LSTM has long been utilized in natural language processes (NLP) [56,57]. Recently, it has been introduced in the remote sensing field for change detection and land cover mapping. In this section, we exploit a bi-directional LSTM [57] to learn the relationship between multi-temporal spatial features extracted from the UAV image. As shown in Figure 5, two LSTMs are stacked together while the hidden state of first LSTM is fed into the second one, and the second LSTM follows a reverse order of the former to fully understand the dependencies of the sequential signals in a bi-directional way. In addition, to further improve the performance, we append an attention layer to the output of the second LSTM. Actually, attention mechanism is widely studied in the field of CV and NLP [58–60], which could automatically adjust the weight of input feature vectors according to their importance to the current task. Therefore, we also incorporate an attention layer to re-weight the sequential features to boost the classification performance.

Let H be a matrix containing a series of vectors $[h_1, h_2, \ldots, h_T]$ that are produced by the bi-directional LSTM, where T denotes the length of the input features. The output of the attention layer is formed by a weighted sum of vectors described as follows:

$$M = \tanh(H) \tag{7}$$

$$\alpha = softmax(w^T M) \tag{8}$$

$$R_{att} = H\alpha^T \tag{9}$$

where α is the attention vector and while R_{att} denotes the fused and attention-weighted spatial-temporal features. Additionally, the features outputted from the Bi-LSTM-Attention are re-weighted or re-calibrated adaptively, which could enhance the informative feature vectors and suppress the noisy and useless ones.

Finally, all the reweighted features were firstly sent to a fully-connected layer and then to a softmax classifier to predict the final vegetable category.

2.6. Details of Network Training

When training started, all the weights of the neural network were initialized through He normalization [61], and biases were all set to be zero. We adopt cross-entropy loss (Equation (10)) [62] as the loss function to train the proposed ARCNN:

$$CE = -\sum_{i} y_i^p \log(y_i) \tag{10}$$

where CE is short for cross-entropy loss, y^p is the predicted result and y is one-hot representation of the ground-truth label. Adam [63] was utilized as the optimization method with a learning rate of 1×10^{-4}. In the training procedure, the model with the lowest validation loss was saved.

We have conducted data augmentation to reduce the impact of limited labeled data in this study. Specifically, all training image patches were flipped and rotated by a random angle from 90°, 180° and 270°. Afterwards, we split 90% of the training datasets for the optimization of parameters. The remaining 10% of the training datasets were utilized as validation sets for performance evaluation during training. After the training process, a testing dataset was adopted to obtain the final classification accuracy.

Furthermore, we used TensorFlow [64] for the construction of our proposed model. The training process was performed on a computer running the Ubuntu 16.04 operation system. The central processing unit (CPU) involved as an Intel core i7-7800 @ 3.5 GHz while the graphics processing unit (GPU) was an NVIDIA GTX TitanX.

2.7. Accuracy Assessment

In this study, we utilized both qualitative and quantitative methods to verify the effectiveness of the proposed ARCNN for vegetable mapping. Specifically, as for the former, we used visual inspection to check for classification errors. While for the latter, a confusion matrix (CM) was obtained from the testing dataset. A series of metrics were calculated from the CM, including overall accuracy (OA), producer's accuracy (PA), user's accuracy (UA) and the Kappa coefficient.

As for the numbers of points chosen for each class, they were actually determined by the area ratio. For instance, the class of Chinese cabbage had the largest area ratio, therefore, the number of training/testing sample points was set to 400, which was the biggest among all the categories. Furthermore, other land cover types, such as spinach and kohlrabi, which only accounted for a small area on the entire study region, had a small number of sample points (only 50).

To further justify the effectiveness of the proposed method, we adopted both ablation analysis and comparison experiments with classic machine learning methods. Specifically, as for the ablation study, we justified the role of both attention-based RNN on vegetable mapping using the following setups. (1) Feature-stacking: concatenating or stacking the spatial features derived from each single-date data for classification; (2) Bi-LSTM: using a bi-directional LSTM for classification; (3) Bi-LSTM-Attention: using the attention-based bi-directional LSTM for classification and (4) standard convolution: using common, non-deformable convolution operations for classification. Besides, ablation study has also been done to justify the impact of deformable convolution when compared with the standard convolution operations.

Meanwhile, classic machine learning methods such as MLC, RF and SVM were also included for comparison experiments. In specific, MLC has long been studied in remote sensing image classification, where the predicted labels are generated based on the maximum likelihood when compared with the training samples. The basic assumption of MLC is that the training samples should follow the normal distribution, which is hard to satisfy in reality, resulting in a limited classification performance. RF belongs to an ensemble of decision trees and the predicted results are determined by the average output of each decision tree [65]. RF has no restrictions on training data distribution and has outperformed MLC in many remote sensing studies. As for SVM, it is based on the Vapnik–Chervonenkis (VC) dimension theory which aims at the minimization of structure risk, resulting in good performance, especially under the situation of limited data [66]. Parameters involved in SVM usually contain kernel type, penalty coefficient, etcetera.

3. Results

3.1. Results of Vegetable Mapping Based on Multi-Temporal Data

Figure 7 shows the vegetable mapping result generated from the proposed ARCNN. It manifests that the distribution of each vegetable category is close to field surveyed when compared with the ground truth (GT) map of Figure 8. The classification errors mainly lie between Chinese cabbage and leaf mustard, potherb, turnip and spinach.

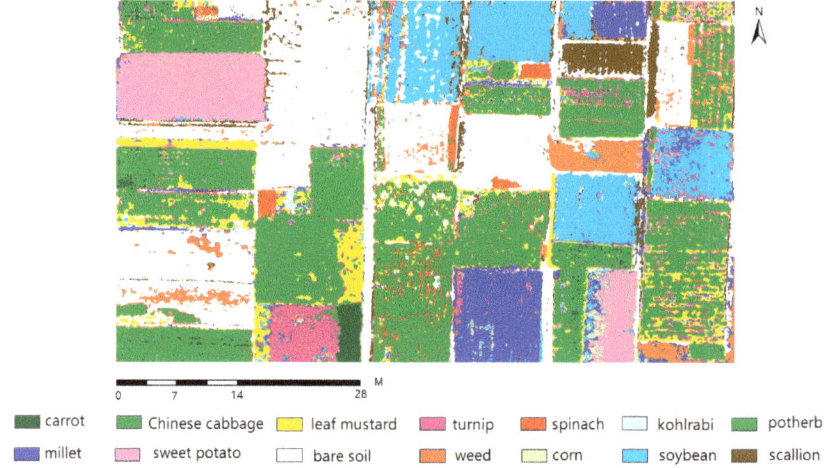

Figure 7. Vegetable map generated from the proposed ARCNN and multi-temporal UAV RGB datasets.

In order to further visually justify the classification results, Figure 8 shows the ground truth map which is manually vectorized from the UAV data. Actually, when compared with the GT map, the classification map of Figure 7 shows a salt and pepper effect. On one side, the classification model in this research belongs to a per-pixel method, which does not consider the boundary information of each land parcel, resulting in a more scattered classification result. On the other hand, the GT map is an ideal description of the spatial extent of every land cover category, neglecting the variations within each land parcel. For instance, several weed regions are missing in the GT map due to small areas. Additionally the bare soil regions in some land parcels have also been neglected. However, in practice, based on the classification map of Figure 7, we could easily generate a land cover map which is more accurate and concise just like Figure 8, which would justify the value of the proposed method in vegetable mapping. To make the boundaries of each land parcel more accurate, in future study we will research semantic segmentation models such as fully convolutional neural networks [19] to improve the visual effect.

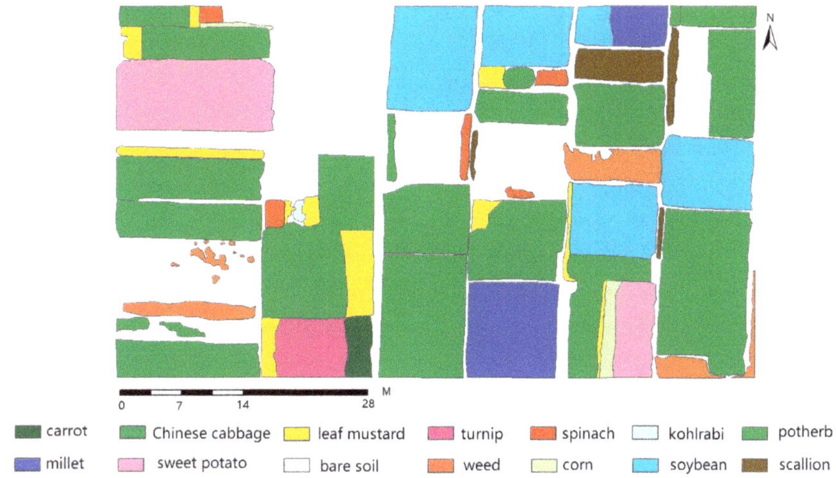

Figure 8. Ground truth map vectorized from the UAV data.

Meanwhile, to quantitatively assess the classification performance, the confusion matrix, Kappa coefficient, OA, PA and UA were derived from the testing dataset. Table 4 indicates that the proposed classification model shows a high performance with both a high OA (92.80%) and a high Kappa coefficient (0.9206).

Table 4. Confusion matrix.

Class	Ground Truth														
	1	2	3	4	5	6	7	8	9	10	11	12	13	14	UA
1	200	9	0	0	0	0	0	0	0	0	0	0	0	0	95.7
2	0	353	21	3	0	0	0	0	0	0	0	0	0	0	93.6
3	0	36	149	12	0	0	0	0	0	0	0	0	0	0	75.6
4	0	0	7	182	0	0	0	16	0	0	0	0	4	0	87.1
5	0	0	0	0	50	0	0	0	0	0	0	0	0	0	100
6	0	0	0	0	0	50	0	0	0	0	0	0	0	0	100
7	0	0	16	0	0	0	100	0	0	0	0	0	0	0	86.2
8	0	0	2	0	0	0	0	184	0	0	0	8	0	0	94.8
9	0	0	0	0	0	0	0	0	100	0	0	0	0	0	100
10	0	0	0	0	0	0	0	0	0	200	0	2	0	9	94.8
11	0	0	0	0	0	0	0	0	0	0	200	0	0	0	200
12	0	0	0	1	0	0	0	0	0	0	0	33	0	0	97.1
13	0	2	5	2	0	0	0	0	0	0	0	7	196	0	92.5
14	0	0	0	0	0	0	0	0	0	0	0	0	0	91	100
PA	100	88.3	74.5	91.0	100	100	100	92.0	100	100	100	66.0	98.0	91.0	
OA		92.80		Kappa							0.9206				

1: carrot; 2: Chinese cabbage; 3: leaf mustard; 4: turnip; 5: spinach; 6: kohlrabi; 7: potherb; 8: millet; 9: weed; 10: bare soil; 11: sweet potato; 12: corn; 13: soybean; 14: scallion; PA: producer's accuracy; UA: user's accuracy and OA: overall accuracy.

Table 4 indicates that the omissions and commissions mainly exist among leaf mustard and Chinese cabbage, potherb and turnip. For instance, several leaf mustard pixels were misclassified as Chinese cabbage and vice versa. This was understandable, since both color and shape of these leafy green vegetables (Chinese cabbage, leaf mustard, potherb, etc.) are very similar, especially at the early growth stage. Meanwhile, the RGB image has the drawback of a low spectral resolution, making it hard to differentiate these vegetable categories when using only color and shape information.

In addition, only a few mistakes occurred among the other categories, which verifies the effectiveness of the proposed vegetation mapping method.

3.2. Results of Vegetable Mapping Based on Mono-Temporal UAV Data

Figure 9 shows the vegetable map generated from both mono- and multi-temporal classification.

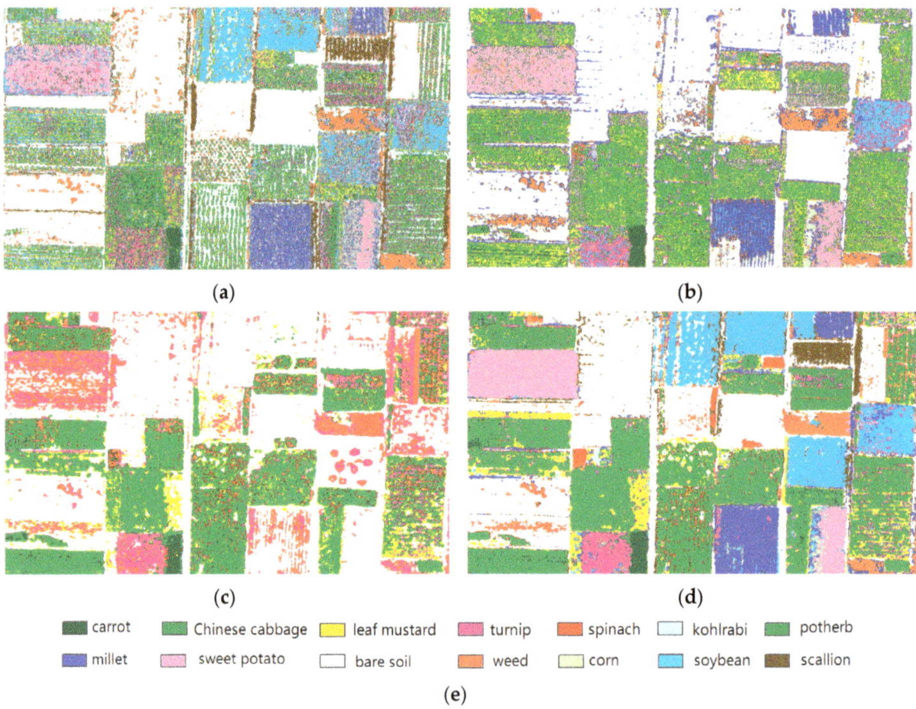

Figure 9. Vegetable map generated from (**a**) T1/2019-09; (**b**) T2/2019-10; (**c**) T3/2019-11 and (**d**) multi-temporal datasets using the proposed method. (**e**) Legend.

As mentioned above, one hypothesis of this study is that the inclusion of multi-temporal UAV data could provide additional phenological information, which would enhance the inter-class separability to cover the shortage of low spectral resolution caused by off-the-shelf digital cameras. Therefore, in this section, a contrast experiment was conducted to compare the performance between multi-temporal and mono-temporal classification. It should be noted that when using single-date UAV data for classification, the spatial-temporal feature fusion module (i.e., Bi-LSTM-Attention) would be non-functional during the training and testing procedure.

Figure 9 indicates that the incorporation of multi-temporal UAV images could significantly improve the classification performance when compared with mono-temporal data, which shows fewer obvious errors from visual inspection. This is in accordance with quantitative assessment (Table 5). It indicates that the overall classification accuracy improved by 19.76%–28.13%, with an average increase of 24.49%, after the inclusion of multi-temporal data.

Table 5. Class-level accuracy for mono- and multi-temporal classification.

No.	Class Name	T1 (%)	T2 (%)	T3 (%)	Proposed (%)
1	Carrot	89.00	93.00	84.00	100
2	Chinese cabbage	46.50	68.50	77.25	88.25
3	Leaf mustard	29.00	39.00	50.00	74.50
4	Turnip	59.00	43.50	75.00	91.00
5	Spinach	–	38.00	18.00	100
6	Kohlrabi	22.00	34.00	–	100
7	Potherb	–	25.00	57.00	100
8	Millet	73.50	76.50	–	92.00
9	Weed	75.00	95.00	85.00	100
10	Bare soil	92.50	94.50	79.00	100
11	Sweet potato	55.50	84.00	–	100
12	Corn	78.00	76.00	–	66.00
13	Soybean	75.00	49.00	–	98.00
14	Scallion	95.00	–	–	91.00
	OA (%)	64.67	67.22	73.04	92.80
	Kappa	0.6067	0.6314	0.6744	0.9206

Meanwhile, Figure 9 also shows that it is difficult to obtain a high-precision vegetable map if only utilizing single-date UAV RGB images. There would be a large amount of classification errors among different vegetable categories, especially between Chinese cabbage, leaf mustard and turnip. Specifically, during the early growth stage (T1), large amounts of Chinese cabbage and leaf mustard pixels are misclassified as turnip (Figure 9a). This is mainly because these leafy green vegetables share very similar appearances (e.g., color, shape and texture patterns), which leads to a low inter-class separability hence a poor classification accuracy (64.67%). In the middle growth stage (T2), the classification accuracy of Chinese cabbage has been greatly improved due to its shape change due to the growth process. However, it still remains difficult to separate leaf mustard from Chinese cabbage (Figure 9b). When it comes to the ripe stage (T3), the leaf mustard could finally be differentiated from Chinese cabbage (Figure 9c). This is mainly because the Chinese cabbage shows a round head in the ripe stage (Table 1), which is greatly different to leaf mustard.

Table 5 shows the class-level accuracy for each vegetable category and other land cover types. It indicates that there is a significant accuracy gap between mono- and multi-temporal classification when using UAV RGB imagery. This is understandable because if using single-date UAV data alone, the similarity of color and texture patterns between various vegetables would yield a low inter-class separability. This is even more so at the early growth stage (T1), when vegetable seedlings share very similar appearances, resulting in the lowest classification accuracy with an OA of 64.67%. However, with the inclusion of multi-temporal UAV images, the additional phenological information would increase the separability among various vegetables, which could boost the final classification performance.

3.3. *Results of Ablation Analysis*

To justify the effectiveness of the proposed ARCNN model, a series of ablation experiments are conducted and the results are shown as follows.

3.3.1. Results of Different Fusion Methods

In this section, we consider the following methods for the fusion of spatial-temporal features: (1) feature-stacking; (2) Bi-LSTM and (3) Bi-LSTM-Attention. The description of these methods is in Section 2.7. The experimental results are shown in Table 6.

Table 6. Comparison between different spatial-temporal feature fusion methods.

Method	OA (%)	Kappa
Feature-stacking	89.56	0.8849
Bi-LSTM	90.93	0.8999
Bi-LSTM-Attention	92.80	0.9206

Table 6 indicates that the Bi-LSTM-Attention module used in this study outperforms both feature-stacking and Bi-LSTM, which increases the OA by 3.24% and 1.87%, respectively. The role of Bi-LSTM-Attention will be discussed in Section 4.1.

3.3.2. Results of Standard Convolution

In this section, we replaced all the deformable convolution operations by standard convolution units in the proposed network to justify the role of deformable convolution in vegetable mapping. Table 7 shows the comparison results.

Table 7. Comparison of standard and deformable convolution.

Method	OA (%)	Kappa
Standard convolution	91.96	0.9111
Deformable convolution	92.80	0.9206

Table 7 implies that the inclusion of deformable convolution could improve the vegetable mapping accuracy. The detailed discussion will be presented in Section 4.2.

3.4. *Results of Comparison with Other Methods*

To further justify the effectiveness of the proposed classification model, we compared it with several machine learning classifiers and other deep learning models. As for the former, we conducted comparison experiments using MLC, RF and SVM based on the same training and testing datasets. We used grid search for the parameterization of both RF and SVM. It turns out that an RF with 300 decision trees and a max depth of 15, and a SVM with radial basis kernel, a gamma [66] of 0.001 and a penalty coefficient (C) [66] of 100 has the best performance, respectively.

Table 8 shows the comparison results between the proposed model and other classical machine learning methods. It indicates that the deep learning based model has an advantage over the classical methods. A detailed discussion of this will follow in Section 4.4.

Table 8. Comparison with classical machine learning classifiers.

Method	OA (%)	Kappa
MLC	46.04	0.4095
RF	63.96	0.6023
SVM	84.76	0.8317
Proposed	92.80	0.9206

In addition, we have conducted comparison experiments with several previous studies, mainly including Ndikumana et al. (stacked LSTMs) [46], Mou et al. (CNN-RNN) [47] and Ji et al. (3D-CNN) [43]. Because the dimension of input and output of these models are different from ours, we have made necessary changes accordingly when reproducing these DL models. The experimental results are shown as follows.

Table 9 indicates that the proposed ARCNN in the research has a better performance when compared with several previous deep learning models. The OA is boosted by an increase of 2.53%

to 8.36% while the Kappa has risen by 2.78% to 9.23%. The detailed discussion will be presented in Section 4.4.

Table 9. Comparison with other deep learning methods.

Method	OA (%)	Kappa
Ndikumana et al.	84.44	0.8283
Mou et al.	90.27	0.8928
Ji et al.	90.18	0.8915
Proposed	92.80	0.9206

4. Discussion

4.1. Impact of Attention-Based RNN on Vegetable Mapping

In this section, we will discuss the impact of attention mechanisms in the RNN for vegetable mapping. Specifically, according to the ablation study results of Section 3.3.1, the comparisons were made between different methods for spatial-temporal feature fusion, including feature-stacking, Bi-LSTM and Bi-LSTM-Attention. Results show that the feature-stacking yields the lowest accuracy with an OA of 89.56% and a Kappa of 0.8849. The reason is that feature-stacking just concatenates all the multi-temporal features without considering the relationship and temporal dependencies across the sequential UAV data. Meanwhile, since Bi-LSTM could understand the dependencies of the sequential features in a bi-directional way, therefore, it shows a better performance than the simple feature-stacking method with an OA improvement of 1.37%. In this study, we added an attention layer on the top of Bi-LSTM to further improve its performance. The attention based Bi-LSTM could enhance the important features while suppressing the less informative ones, outperforming both feature-stacking and Bi-LSTM with an OA increase of 3.24% and 1.87%, respectively, which verifies its effectiveness in spatial-temporal feature fusion.

4.2. Impact of Deformable Convolution on Vegetable Mapping

Another hypothesis of this study is that the scale and shape variations could be accounted for using the deformable convolution. According to Table 7, it indicates that the inclusion of deformable convolution could boost the classification performance. The OA has been improved from 91.96% to 92.80% with a rise of about 1%, justifying the role of deformable convolution. The reason for the lower accuracy of standard convolution is that it has a fixed kernel shape, which lacks the capability to model the geometric transformations of complex landscapes. On the other hand, deformable convolution has a flexible receptive field, which could be adaptive to the variability of shape and scale of remotely sensed imagery [41]. Therefore, the deformable convolution shows a better performance, especially under the complex and fragmented agricultural landscape in this study.

4.3. Impact of Multi-Temporal UAV Data on Vegetable Mapping

In addition, we will further discuss the role of multi-temporal UAV data on vegetable mapping. In fact, one of the main objectives of this study is to explore whether the incorporation of multi-temporal UAV RGB images could improve the vegetable classification accuracy. The initial motivation lies in the fact that RGB images acquired by UAV have a low spectral resolution, which would make it hard for the fine-grained classification of various vegetable categories. Therefore, in this study we have selected images from three important periods, i.e., the sowing period, the growing period and the harvesting period, to capture the phenological characteristics of different vegetables. Although the number of three dates may seem limited, all of them fall into the distinct periods of the vegetable and crop growth stage, which could still provide additional and useful time-series features for classification.

Meanwhile, in previous studies, images from only three dates have been studied for remote sensing image classification and they outperform the single-date dataset. For instance, Palchowdhuri et al. used three images from both multi-temporal Sentinel-2 and WordView-3 imagery for crop classification in Coalville in the United Kingdom and achieved an accuracy of 91% [67]. Similar findings were also reported in Yang et al., where three images from summer, autumn and winter were integrated for coastal land cover mapping [68]. In our previous study [15], we also utilized only three images during the whole crop growing season for the cropland classification in the Yellow River Delta of China, which yielded an average accuracy of 89%, justifying the role of three dates for classification. In future research, images from a longer temporal range could be included to further improve the classification performance.

4.4. Comparison with Other Methods

In this section, we focused on the detailed discussion between the proposed ARCNN and other classical machine learning methods and several previous deep learning methods. Specifically, Table 8 indicates that our proposed method outperforms machine learning methods such as MLC, RF and SVM with an OA increase of 27.29%, 21.58% and 8.64%, respectively. The results are in accordance with [46] and our previous studies [39,44]. The reason could be that classical machine learning methods lack the ability to capture the high-level representative features when compared to deep learning models, leading to a performance gap in vegetable mapping.

In addition, there is a need to compare the proposed ARCNN with other methods for multi-temporal UAV image classification. Recent researches such as van Iersel et al. [14] and Michez et al. [12], they both utilized object-oriented image analysis (OBIA) and random forest for plant classification from multi-date UAV data. Manually designed features such as band ratio and vegetation indices were used for classification. Compared with their studies, we replace the manually designed features with high-level and discriminative features that are automatically learned from deep neural networks, (i.e., CNN and RNN), which could enhance the feature's representativeness. To the best of our knowledge, this study is the first case to introduce deep learning methods in multi-temporal UAV image classification. Therefore, the proposed method in this research might provide useful reference for future studies.

Meanwhile, it is also necessary to compare the proposed ARCNN with other deep learning models for remote sensing image classification. Early studies mainly utilized LSTM for multi-temporal classification. One representative research is Ndikumana et al., where five LSTMs were stacked for the classification using multi-temporal SAR Sentinel-1 data [46]. The input data in Ndikumana's study are a single pixel with a time curve, which neglects the rich, contextual relationship hidden in the spatial features, showing a relatively lower accuracy (84.44%). Different from Ndikumana et al. [46], we have added a CNN in front of LSTM to enrich the representative spatial feature extraction.

Mou et al. also cascaded a CNN and RNN for change detection from two optical remote sensing images [47]. Compared with Mou et al. [47], our model makes two significant improvements. Firstly, from the perspective of CNN, we incorporate the multi-scale deformable convolutions, which could aggregate multi-level contextual features. Secondly, we used the attention mechanism with a bi-directional LSTM to further enhance the modeling of sequential signals in multi-temporal remote sensing data. All the above modifications have improved the classification from 90.18% by Mou et al. to 92.80%.

Besides, Ji et al. adopted a 3D-CNN to extract spatial-temporal features for crop type mapping from multi-temporal satellite imagery [43]. Compared with Ji et al. [43], our method has also gained a more accurate result. The reason lies in that 3D CNN cannot explicitly establish the relationship between the sequential signals, which has flaws in the generation of spatial-temporal feature fusion and integration. Furthermore, our Bi-LSTM-Attention module is more straightforward in mining the relationship across multi-temporal data than a 3D-CNN.

5. Conclusions

This study proposed an attention-based recurrent convolutional neural network (ARCNN) for accurate vegetable mapping based on multi-temporal unmanned aerial vehicle (UAV) red-green-blue

(RGB) data. The proposed ARCNN first leverages a multi-scale deformable CNN to learn and extract the rich spatial features from each mono-temporal UAV image, which aims to account for the shape and scale variations under complex and fragmented agricultural landscapes. Afterwards, an attention-based bi-directional long-short term memory (LSTM) is introduced to model the relationship between the sequential features, from which spatial and temporal features are fused and aggregated. Finally, the fused features are fed to a fully connected layer and a softmax classifier to determine the vegetable category.

Experimental results showed that the proposed ARCNN yields a high classification performance with an overall accuracy (OA) of 92.08% and a Kappa coefficient of 0.9206. When compared with mono-temporal classification, the incorporation of multi-temporal UAV data could boost the OA significantly by an average increase of 24.49%, which verifies the hypothesis that multi-temporal UAV observations could enhance the inter-class separability and thus reduce the drawback of low spectral resolution of off-the-shelf digital cameras. The Bi-LSTM-Attention module outperforms other fusion methods such as feature-staking and bi-directional LSTM with an OA increase of 3.24% and 1.87%, respectively, justifying its effectiveness in modeling the dependency across the sequential features. Meanwhile, the introduction of deformable convolution could also improve the OA by about 1% when compared with standard convolution. In addition, the proposed ARCNN also shows a higher performance than other classical machine learning classifiers such as maximum likelihood classifier, random forest and support vector machine, and several previous deep learning methods for remote sensing classification.

This study demonstrates that the proposed ARCNN could yield an accurate vegetable mapping result from multi-temporal UAV RGB data. The drawback of low spectral resolution of RGB images could be compensated by introducing additional phenological information and robust deep learning models. Although images from only three dates were included, a good classification result could still be achieved providing all three dates fall into the distinct growing periods of vegetables. Finally, the proposed model could be viewed as a general framework for multi-temporal remote sensing image classification. As for future work, more study cases should be considered to justify the effectiveness of the proposed method. Additionally, semantic segmentation models should be incorporated to get a more accurate vegetable map.

Author Contributions: Methodology, Q.F.; validation, Q.F. and J.Y.; data curation, Y.L. and C.O.; writing—original draft preparation, Q.F.; writing—review and editing, J.Y., D.Z., B.N., J.L. and B.L. All authors have read and agreed to the published version of the manuscript.

Funding: This research was funded by the China Postdoctoral Science Foundation, grant number 2018M641529, 2019T120155 and the National Key Research and Development Program of China, grant number 2018YFE0122700.

Acknowledgments: Special thanks to the anonymous reviewers and editors for their very useful comments and suggestions to help improve the quality of this paper. The authors give thanks to Zhe Liu from the China Agricultural University for assisting in the field work. We would also like to thank Beijing IRIS Remote Sensing Technology Limited, Inc. for their help in preprocessing UAV raw data.

Conflicts of Interest: The authors declare no conflict of interest.

References

1. Torres-Sánchez, T.; Peña, J.M.; de Castro, A.I.; López-Granados, F. Multi-temporal mapping of the vegetation fraction in early-season wheat fields using images from UAV. *Comput. Electron. Agric.* **2014**, *103*, 104–113. [CrossRef]
2. Wikantika, K.; Uchida, S.; Yamamoto, S. Mapping vegetable area with spectral mixture analysis of the Landsat-ETM. In Proceedings of the IEEE International Geoscience and Remote Sensing Symposium, Toronto, ON, Canada, 24–28 June 2002; pp. 1965–1967.
3. Belgiu, M.; Csillik, O. Sentinel-2 cropland mapping using pixel-based and object-based time-weighted dynamic time warping analysis. *Remote Sens. Environ.* **2018**, *204*, 509–523. [CrossRef]

4. Rupasinghe, P.A.; Milas, A.S.; Arend, K.; Simonson, M.A.; Mayer, C.; Mackey, S. Classification of shoreline vegetation in the Western Basin of Lake Erie using airborne hyperspectral imager HSI2, Pleiades and UAV data. *Int. J. Remote Sens.* **2019**, *40*, 3008–3028. [CrossRef]
5. Wan, S.; Chang, S.H. Crop classification with WorldView-2 imagery using Support Vector Machine comparing texture analysis approaches and grey relational analysis in Jianan Plain, Taiwan. *Int. J. Remote Sens.* **2019**, *40*, 8076–8092. [CrossRef]
6. Asgarian, A.; Soffianian, A.; Pourmanafi, S. Crop type mapping in a highly fragmented and heterogeneous agricultural landscape: A case of central Iran using multi-temporal Landsat 8 imagery. *Comput. Electron. Agric.* **2016**, *127*, 531–540. [CrossRef]
7. Feng, Q.; Liu, J.; Gong, J. UAV remote sensing for urban vegetation mapping using random forest and texture analysis. *Remote Sens.* **2015**, *7*, 1074–1094. [CrossRef]
8. Feng, Q.; Liu, J.; Gong, J. Urban flood mapping based on unmanned aerial vehicle remote sensing and random forest classifier—A case of Yuyao, China. *Water* **2015**, *7*, 1437–1455. [CrossRef]
9. Dai, Y.; Gong, J.; Li, Y.; Feng, Q. Building segmentation and outline extraction from UAV image-derived point clouds by a line growing algorithm. *Int. J. Digit. Earth* **2017**, *10*, 1077–1097. [CrossRef]
10. Böhler, J.E.; Schaepman, M.E.; Kneubühler, M. Optimal timing assessment for crop separation using multispectral unmanned aerial vehicle (UAV) data and textural features. *Remote Sens.* **2019**, *11*, 1780. [CrossRef]
11. Pádua, L.; Marques, P.; Hruška, J.; Adão, T.; Peres, E.; Morais, R.; Sousa, J.J. Multi-temporal vineyard monitoring through UAV-based RGB imagery. *Remote Sens.* **2018**, *10*, 1907. [CrossRef]
12. Michez, A.; Piégay, H.; Lisein, J.; Claessens, H.; Lejeune, P. Classification of riparian forest species and health condition using multi-temporal and hyperspatial imagery from unmanned aerial system. *Environ. Monit. Assess.* **2016**, *188*, 146. [CrossRef] [PubMed]
13. Moeckel, T.; Dayananda, S.; Nidamanuri, R.R.; Nautiyal, S.; Hanumaiah, N.; Buerkert, A.; Wachendorf, M. Estimation of vegetable crop parameter by multi-temporal UAV-borne images. *Remote Sens.* **2018**, *10*, 805. [CrossRef]
14. Van Iersel, W.; Straatsma, M.; Middelkoop, H.; Addink, E. Multitemporal Classification of river floodplain vegetation using time series of UAV images. *Remote Sens.* **2018**, *10*, 1144. [CrossRef]
15. Feng, Q.; Gong, J.; Liu, J.; Li, Y. Monitoring cropland dynamics of the yellow river delta based on multi-temporal Landsat imagery over 1986 to 2015. *Sustainability* **2015**, *7*, 14834–14858. [CrossRef]
16. Chen, L.; Yang, W.; Xu, K.; Xu, T. Evaluation of local features for scene classification using VHR satellite images. In Proceedings of the 2011 Joint Urban Remote Sensing Event, Munich, Germany, 11–13 April 2011; pp. 385–388.
17. Zhu, Q.; Zhong, Y.; Zhao, B.; Xia, G.; Zhang, L. Bag-of-visual-words scene classifier with local and global features for high spatial resolution remote sensing imagery. *IEEE Geosci. Remote Sens. Lett.* **2016**, *13*, 747–751. [CrossRef]
18. LeCun, Y.; Bengio, Y.; Hinton, G. Deep learning. *Nature* **2015**, *521*, 436–444. [CrossRef]
19. Shelhamer, E.; Long, J.; Darrell, T. Fully convolutional networks for semantic segmentation. *IEEE Trans. Pattern Anal. Mach. Intell.* **2017**, *39*, 640–651. [CrossRef]
20. LaLonde, R.; Bagci, U. Capsules for Object Segmentation. *arXiv* **2018**, arXiv:1804.04241. Available online: https://arxiv.org/abs/1804.04241.pdf (accessed on 17 April 2020).
21. Lin, T.Y.; Goyal, P.; Girshick, R.; He, K.; Dollar, P. Focal loss for dense object detection. In Proceedings of the IEEE International Conference on Computer Vision (ICCV), Venice, Italy, 22–29 October 2017; pp. 2999–3007.
22. Krizhevsky, A.; Sutskever, I.; Hinton, G.E. Imagenet classification with deep convolutional neural networks. *Proc. Adv. Neural Inf. Process. Syst.* **2012**, 1097–1105. [CrossRef]
23. He, K.; Zhang, X.; Ren, S.; Sun, J. Deep residual learning for image recognition. In Proceedings of the IEEE Conference on Computer Vision and Pattern Recognition, Las Vegas, NV, USA, 27–30 June 2016; pp. 770–778.
24. Hu, J.; Shen, L.; Sun, G. Squeeze-and-Excitation Networks. *arXiv* **2017**, arXiv:1709.01507. Available online: https://arxiv.org/pdf/1709.01507.pdf (accessed on 17 April 2020).
25. Zhu, X.X.; Tuia, D.; Mou, L.; Xia, G.; Zhang, L.; Xu, F.; Fraundorfer, F. Deep learning in remote sensing: A comprehensive review and list of resources. *IEEE Geosci. Remote Sens. Mag.* **2017**, *5*, 8–36. [CrossRef]
26. Zhang, L.; Zhang, L.; Du, B. Deep learning for remote sensing data: A technical tutorial on the state of the art. *IEEE Geosci. Remote Sens. Mag.* **2016**, *4*, 22–40. [CrossRef]

27. Kellenberger, B.; Marcos, D.; Tuia, D. Detecting mammals in UAV images: Best practices to address a substantially imbalanced dataset with deep learning. *Remote Sens. Environ.* **2018**, *216*, 139–153. [CrossRef]
28. Carrio, A.; Sampedro, C.; Rodriguez-Ramos, A.; Campoy, P. A review of deep learning methods and applications for unmanned aerial vehicles. *J. Sens.* **2017**, 3296874. [CrossRef]
29. Kamilaris, A.; Prenafeta-Boldú, F.X. Deep learning in agriculture: A survey. *Comput. Electron. Agric.* **2018**, *147*, 70–90. [CrossRef]
30. Chen, Y.; Fan, R.; Bilal, M.; Yang, X.; Wang, J.; Li, W. Multilevel cloud detection for high-resolution remote sensing imagery using multiple convolutional neural networks. *ISPRS Int. J. Geo. Inf.* **2018**, *7*, 181. [CrossRef]
31. Alshehhi, R.; Marpu, P.R.; Woon, W.L.; Mura, M.D. Simultaneous extraction of roads and buildings in remote sensing imagery with convolutional neural networks. *ISPRS J. Photogramm. Remote Sens.* **2017**, *130*, 139–149. [CrossRef]
32. Xu, Y.; Wu, L.; Xie, Z.; Chen, Z. Building Extraction in Very High Resolution Remote Sensing Imagery Using Deep Learning and Guided Filters. *Remote Sens.* **2018**, *10*, 144. [CrossRef]
33. Li, X.; Yao, X.; Fang, Y. Building-A-Nets: Robust Building Extraction From High-Resolution Remote Sensing Images With Adversarial Networks. *IEEE J. Sel. Top. Appl. Earth Obs. Remote Sens.* **2018**, *11*, 3680–3687. [CrossRef]
34. Deng, Z.; Sun, H.; Zhou, S.; Zhao, J.; Zou, H. Toward Fast and Accurate Vehicle Detection in Aerial Images Using Coupled Region-Based Convolutional Neural Networks. *IEEE J. Sel. Top. Appl. Earth Obs. Remote Sens.* **2017**, *10*, 3652–3664. [CrossRef]
35. Ammour, N.; Alhichri, H.; Bazi, Y.; Benjdira, B.; Alajlan, N.; Zuair, M. Deep Learning Approach for Car Detection in UAV Imagery. *Remote Sens.* **2017**, *9*, 312. [CrossRef]
36. Han, W.; Feng, R.; Wang, L.; Cheng, Y. A semi-supervised generative framework with deep learning features for high-resolution remote sensing image scene classification. *ISPRS J. Photogramm. Remote Sens.* **2018**, *145*, 23–43. [CrossRef]
37. Wang, Q.; Liu, S.; Chanussot, J.; Li, X. Scene classification with recurrent attention of VHR remote sensing images. *IEEE Trans. Geosci. Remote Sens.* **2019**, *57*, 1155–1167. [CrossRef]
38. Rußwurm, M.; Körner, M. Multi-temporal land cover classification with sequential recurrent encoders. *ISPRS Int. J. Geo. Inf.* **2018**, *7*, 129. [CrossRef]
39. Feng, Q.; Zhu, D.; Yang, J.; Li, B. Multisource hyperspectral and LiDAR data fusion for urban land-use mapping based on a modified two-branch convolutional neural network. *ISPRS Int. J. Geo. Inf.* **2019**, *8*, 28. [CrossRef]
40. Huang, B.; Zhao, B.; Song, Y. Urban land-use mapping using a deep convolutional neural network with high spatial resolution multispectral remote sensing imagery. *Remote Sens. Environ.* **2018**, *214*, 73–86. [CrossRef]
41. Kussul, N.; Lavreniuk, M.; Skakun, S.; Shelestov, A. Deep learning classification of land cover and crop types using remote sensing data. *IEEE Geosci. Remote Sens. Lett.* **2017**, *14*, 778–782. [CrossRef]
42. Rezaee, M.; Mahdianpari, M.; Zhang, Y.; Salehi, B. Deep convolutional neural network for complex wetland classification using optical remote sensing imagery. *IEEE J. Sel. Top. Appl. Earth Obs. Remote Sens.* **2018**, *11*, 3030–3039. [CrossRef]
43. Ji, S.; Zhang, C.; Xu, A.; Shi, Y.; Duan, Y. 3D convolutional neural networks for crop classification with multi-temporal remote sensing images. *Remote Sens.* **2018**, *10*, 75. [CrossRef]
44. Feng, Q.; Yang, J.; Zhu, D.; Liu, J.; Guo, H.; Bayartungalag, B.; Li, B. Integrating multitemporal sentinel-1/2 data for coastal land cover classification using a multibranch convolutional neural network: A case of the Yellow River Delta. *Remote Sens.* **2019**, *11*, 1006. [CrossRef]
45. Hochreiter, S.; Schmidhuber, J. Long short-term memory. *Neural Comput.* **1997**, *9*, 1735–1780. [CrossRef] [PubMed]
46. Ndikumana, E.; Ndikumana, E.; Ho Tong Minh, D.; Baghdadi, N.; Courault, D.; Hossard, L. Deep recurrent neural network for agricultural classification using multitemporal SAR Sentinel-1 for Camargue, France. *Remote Sens.* **2018**, *10*, 1217. [CrossRef]
47. Mou, L.; Bruzzone, L.; Zhu, X.X. Learning spectral-spatial-temporal features via a recurrent convolutional neural network for change detection in multispectral imagery. *IEEE Trans. Geosci. Remote Sens.* **2019**, *57*, 924–935. [CrossRef]
48. Song, A.; Choi, J.; Han, Y.; Kim, Y. Change detection in hyperspectral images using recurrent 3D fully convolutional networks. *Remote Sens.* **2018**, *10*, 1827. [CrossRef]

49. Liu, Q.; Zhou, F.; Hang, R.; Yuan, X. Bidirectional-convolutional LSTM based spectral-spatial feature learning for hyperspectral image classification. *Remote Sens.* **2017**, *9*, 1330.
50. Mou, L.; Ghamisi, P.; Zhu, X.X. Deep recurrent neural networks for hyperspectral image classification. *IEEE Trans. Geosci. Remote Sens.* **2017**, *55*, 3639–3655. [CrossRef]
51. DJI-Inspire 2. Available online: https://www.dji.com/cn/inspire-2/ (accessed on 17 March 2020).
52. Pix4D. Available online: http://pix4d.com/ (accessed on 17 March 2020).
53. ENVI. Available online: http://www.enviidl.com/ (accessed on 17 March 2020).
54. Dai, J.; Qi, H.; Xiong, Y.; Li, Y.; Zhang, G.; Hu, H.; Wei, Y. Deformable convolutional networks. *arXiv* **2017**, arXiv:1703.06211. Available online: https://arxiv.org/abs/1703.06211 (accessed on 17 March 2020).
55. Jin, Q.; Meng, Z.; Pham, T.D.; Chen, Q.; Wei, L.; Su, R. DUNet: A Deformable Network for Retinal Vessel Segmentation. *arXiv* **2018**, arXiv:1811.01206. Available online: https://arxiv.org/pdf/1811.01206.pdf (accessed on 17 April 2020). [CrossRef]
56. Pan, D.; Yuan, J.; Li, L.; Sheng, D. Deep neural network-based classification model for Sentiment Analysis. *arXiv* **2019**, arXiv:1907.02046. Available online: https://arxiv.org/abs/1907.02046 (accessed on 17 April 2019).
57. Melamud, O.; Goldberger, J.; Dagan, I. Context2vec: Learning generic context embedding with bidirectional LSTM. In Proceedings of the 20th SIGNLL Conference on Computational Natural Language Learning (CoNLL), Berlin, Germany, 11–12 August 2016; pp. 51–61.
58. Cui, W.; Wang, F.; He, X.; Zhang, D.; Xu, X.; Yao, M.; Wang, Z.; Huang, J. Multi-scale semantic segmentation and spatial relationship recognition of remote sensing images based on an attention model. *Remote Sens.* **2019**, *11*, 1044. [CrossRef]
59. Xu, R.; Tao, Y.; Lu, Z.; Zhong, Y. Attention-mechanism-containing neural networks for high-resolution remote sensing image classification. *Remote Sens.* **2018**, *10*, 1602. [CrossRef]
60. Zhao, B.; Wu, X.; Feng, J.; Peng, Q.; Yan, S. Diversified visual attention networks for fine-grained object classification. *IEEE Trans. Multimedia* **2017**, *19*, 1245–1256. [CrossRef]
61. He, K.; Zhang, X.; Ren, S.; Sun, J. Delving deep into rectifiers: Surpassing human-level performance on ImageNet classification. *arXiv* **2015**, arXiv:1502.01852. Available online: https://arxiv.org/pdf/1502.01852.pdf (accessed on 17 March 2020).
62. Cox, D. The Regression Analysis of Binary Sequences. *J. Royal Stat. Soc. Ser. B* **1958**, *20*, 215–242. [CrossRef]
63. Kingma, D.P.; Ba, J. Adam: A Method for Stochastic Optimization. *arXiv* **2014**, arXiv:1412.6980. Available online: https://arxiv.org/abs/1412.6980 (accessed on 17 March 2020).
64. TensorFlow. Available online: https://tensorflow.google.cn/ (accessed on 17 March 2020).
65. Breiman, L. Random forests. *Mach. Learn.* **2001**, *45*, 5–32. [CrossRef]
66. Chapelle, O.; Vapnik, V.; Bousquet, O.; Mukherjee, S. Choosing multiple parameters for support vector machines. *Mach. Learn.* **2002**, *46*, 131–159. [CrossRef]
67. Palchowdhuri, Y.; Valcarce-Diñeiro, R.; King, P.; Sanabria-Soto, M. Classification of multi-temporal spectral indices for crop type mapping: A case study in Coalville, UK. *J. Agric. Sci.* **2018**, *156*, 24–36. [CrossRef]
68. Yang, X.; Chen, L.; Li, Y.; Xi, W.; Chen, L. Rule-based land use/land cover classification in coastal areas using seasonal remote sensing imagery: A case study from Lianyungang City, China. *Environ. Monit. Assess.* **2015**, *187*, 449. [CrossRef]

Article

Crop Row Detection through UAV Surveys to Optimize On-Farm Irrigation Management

Giulia Ronchetti [1,*], Alice Mayer [2], Arianna Facchi [2], Bianca Ortuani [2] and Giovanna Sona [1]

1 Department of Civil and Environmental Engineering, Politecnico di Milano, 20133 Milan, Italy; giovanna.sona@polimi.it
2 Department of Agricultural and Environmental Sciences – Production, Landscape, Agroenergy, Università degli Studi di Milano, 20133 Milan, Italy; alice.mayer@unimi.it (A.M.); arianna.facchi@unimi.it (A.F.); bianca.ortuani@unimi.it (B.O.)
* Correspondence: giulia.ronchetti@polimi.it; Tel.: +39-0223996543

Received: 11 May 2020; Accepted: 16 June 2020; Published: 18 June 2020

Abstract: Climate change and competition among water users are increasingly leading to a reduction of water availability for irrigation; at the same time, traditionally non-irrigated crops require irrigation to achieve high quality standards. In the context of precision agriculture, particular attention is given to the optimization of on-farm irrigation management, based on the knowledge of within-field variability of crop and soil properties, to increase crop yield quality and ensure an efficient water use. Unmanned Aerial Vehicle (UAV) imagery is used in precision agriculture to monitor crop variability, but in the case of row-crops, image post-processing is required to separate crop rows from soil background and weeds. This study focuses on the crop row detection and extraction from images acquired through a UAV during the cropping season of 2018. Thresholding algorithms, classification algorithms, and Bayesian segmentation are tested and compared on three different crop types, namely grapevine, pear, and tomato, for analyzing the suitability of these methods with respect to the characteristics of each crop. The obtained results are promising, with overall accuracy greater than 90% and producer's accuracy over 85% for the class "crop canopy". The methods' performances vary according to the crop types, input data, and parameters used. Some important outcomes can be pointed out from our study: NIR information does not give any particular added value, and RGB sensors should be preferred to identify crop rows; the presence of shadows in the inter-row distances may affect crop detection on vineyards. Finally, the best methodologies to be adopted for practical applications are discussed.

Keywords: UAV; crop row; precision agriculture; DEM; RGB; vegetation indices

1. Introduction

According to the most recent projections presented by the Intergovernmental Panel on Climate Change (IPCC), the variation in precipitation is altering hydrological systems in many agricultural areas of the planet, affecting water resources in terms of both quantity and quality [1]. In particular, climate change is leading to more frequent extreme events (droughts and heat waves, occurring more often and lasting longer), to the alteration of spatial and temporal precipitation (less precipitation concentrated in a few heavy rainfall events), as well as to an increase in air temperatures and crop water needs in many geographical areas.

In Mediterranean countries, freshwater resources are currently highly exploited due to the rapid population growth and intensive water use in agriculture, industry, and tourism activities. In many areas of Europe, including Italy, the effects of climate change are already detectable and have led to the need to irrigate crops and areas traditionally not irrigated [2].

In the context of Precision Agriculture (PAg), particular attention is given to the optimization of on-farm irrigation management, since water resources for agricultural use have become scarcer due to the combined effect of climate change and the increased competition among different water uses [1,3]. Moreover, due to the intensification of extreme weather events, irrigation is becoming an important tool to guarantee adequate quality standards to agricultural products [4]. The optimization of on-farm irrigation management through variable rate irrigation systems is based on the detection of the spatial variability of soil and crop properties. Variable rate irrigation is aimed at managing water inputs to match the spatial variability of the water requirements found in the field, by providing irrigation in different amounts depending on real crop requirements. This approach represents a valid solution to increase water use efficiency and water savings, and for certain crops, variable water application might lead also to an increase of yield and product quality [5–7].

Currently, the detection of within-field spatial variability of crops can be easily carried out through many different technologies, including proximal and remote sensing techniques, sensors mounted on agricultural machinery, georeferencing systems, and geographical information systems [8]. Among them, the noteworthy development of Unmanned Aerial Vehicles (UAVs) registered in the last few years has allowed filling the gap between remote sensing and terrestrial techniques in agricultural applications [9]. Using UAVs is a good compromise between the large coverage obtainable with remote platforms (mainly satellite and aircraft) and the accuracy of the terrestrial data, with advantages in terms of time consumption and the costs of the surveys.

By allowing collecting information closely from above the crop canopy, UAVs have introduced a new point of view for agricultural surveys, giving rise to several applications. As early as 2008, Nebiker et al. [10] proposed the prototype of a multispectral sensor suitable to be mounted on a UAV and conducted experiments with it to assess vegetation health with promising results. Starting from this experience, a variety of studies can be found in the literature about the effectiveness of UAV surveys conducted for PAg purposes. The main applications involve in-field weed mapping, vegetation growth monitoring, crop water stress analysis, and optimization of irrigation management [9,11].

The major strength of UAV surveys is to provide information in a rapid, non-destructive way with a high spatial resolution that can detect within-field variation in detail. Nevertheless, when UAV imagery is used to monitor crop variability in row-crops, a post-processing procedure must be set up to identify and extract crop rows from soil background and weeds, thus generating a crop mask. This issue is crucial to manage precision irrigation properly, by adjusting water supplies according to crop water requirements, considering the actual crop water status and vigor and their spatial variability. Moreover, the crop row detection is crucial to reliably assess the effectiveness of variable rate irrigation in optimizing the crop growth by reducing crop variability. In both of these cases, a non-accurate recognition of crop rows could induce misleading assessment of crop status, then erroneous evaluation of water amounts to be applied through irrigation, as well as erroneous validation of irrigation management. The same highlights are also valid for fertilization practices based on prescription maps, obtained from NDVI maps acquired through UAV surveys. Fertilization is more effective the more accurate the crop row detection is. As a matter of fact, what has been reported is valid in fruit-viticulture and horticulture, where nutrients are applied by fertigation through drip irrigation systems. On the other hand, crop row detection can be useful even for herbaceous crops sown in rows, such as maize. In the case of maize, procedures to generate crop masks are fundamental to manage precision fertilization or to control weeds in the early growth stages, before the canopy closure when the crop rows are still recognizable [12,13].

Different studies can be found in the literature, proposing (semi)automatic methods, using image-processing techniques on single-band images, maps of Vegetation Indices (VIs), or Digital Elevation Models (DEMs), to detect crop canopy [14]. Poblete-Echeverría et al. [15] compared the performance of four classification methods, including standard and well known methods (i.e., K-means and VIs' thresholding) and machine learning methods (i.e., artificial neural networks and random forest), to detect vine canopy using ultra-high-resolution RGB imagery acquired with a conventional camera mounted on a low-cost UAV. Marques et al. [16] presented a UAV-based automatic method to detect chestnut trees, by using RGB and CIR (Color Infrared) orthomosaics combined with the canopy height model. In [17], potato plant objects were extracted from bare soil using the excess green index and Otsu thresholding methods.

This study focuses on the crop row detection and extraction by analyzing and post-processing images acquired through a UAV. Different methodologies are tested and compared, as well as several segmentation methods, such as supervised classifications, Bayesian segmentation, and thresholding algorithms, developed ad hoc for this purpose. Extraction algorithms are applied both on geometric products (i.e., digital elevation model) and vegetation indices' maps. As already presented by other studies [15,16,18], the proposed methods exploit existing indicators, including NDVI and RGB derived indices. In this way, this study aims at demonstrating the importance and effectiveness of UAV systems in precision agriculture and providing to less experienced users the opportunity to use them. As an added value, using simple tools and proving their usefulness would allow the diffusion of these methodologies to a wide audience, even outside the academic environment. All the described methods are semi-automatic and require little human intervention for parameter choosing and fine-tuning. The assessment of the methods was performed on three different crop types, grapevine, pear, and tomato, to analyze the suitability of the proposed methods with respect to the characteristics of each crop. Considering the application of the proposed algorithms to different agro-ecosystems, each with its own peculiarities, represents the real novelty of this study with respect to the existing literature.

2. Materials

2.1. Study Sites

This study is part of the two year project NUTRIPRECISO (RDP-EU, Measure 1.2.01, Lombardy Region), which involved the Department of Agricultural and Environmental Sciences (DiSAA), University of Milan, the Department of Civil and Environmental Engineering (DICA), Politecnico di Milano, and CREA (Consiglio per la ricerca in agricoltura e l'analisi dell'economia agraria). The project was aimed at developing and disseminating practices of PAg in fruit viticulture and horticulture. Consequently, in this study, simple and reliable methodologies, moreover using commonly available data, are presented to be proposed to the farmers and potential users.

Three different types of crops were investigated in the NUTRIPRECISO project: grapevine, pear, and tomato; therefore, three different areas in the Lombardy Region were chosen as study sites, to be representative of each crop. The location of the vineyard is shown in Figure 1a, while the pear orchard and tomato sites are reported in Figure 1b. In the following, the main characteristics of the three study sites are described.

2.1.1. Vineyard

The vineyard, with an extent of 1 ha, is located in Olfino di Monzambano, in the province of Mantova, the heart of the Morainic Hills region (Northern Italy).
The vineyard is nearly flat and located at an altitude of about 88 m a.s.l. A soil survey with an Electro-Magnetic Induction (EMI) sensor followed by a traditional pedological survey showed that the field, despite its small dimension, was characterized by four different soil types with coarse soils in the northwestern part and fine heavy soils in the eastern part [19].

According to the ARPA (Regional Environmental Protection Agency) agro-meteorological station located at Ponti sul Mincio (about 6.5 km from the vineyard), the climatic conditions are warm and mild with average maximum and minimum temperatures during summer of about 30 and 20 °C, respectively, and rainfall concentrated in spring and autumn, with a mean annual value of 765 mm (values calculated as the average over the period 1993–2017), which provide favorable conditions for the grapevine to grow.

The vineyard variety was Chardonnay, cultivated in rows oriented along the east-west axis, with a distance of plants on the row of 0.8 m and a distance between the rows of 2.4 m, while the row height was about 1.3 m. The plant cover fraction in the phase of maximum development of the canopy was estimated to be about 25% (equivalent to a row width of 0.6 m). The soil, both under the rows and between the rows, was grass-covered with periodic mowing to regulate the excessive development of vegetation [19].

(a) (b)

Figure 1. The experimental sites: (**a**) the vineyard; (**b**) in yellow the pear orchard and in red the tomato field. Coordinate Reference System (CRS): WGS84/UTM Zone 32N. Map data: ©OpenStreetMap contributors.

2.1.2. Pear Orchard

The pear orchard was situated in the "Dotti" farm, a research facility of the University of Milan, located in Arcagna locality, Montanaso Lombardo (province of Lodi (LO)).

The orchard covers an area of about 1 ha and is flat, positioned at about 80 m a.s.l. According to the data recorded in the period 1993–2017 at the nearest ARPA agro-meteorological station (located at about 12 km southeast from the experimental site in Cavenago d'Adda), the climate is characterized by two rainy periods, respectively in April and September, while the highest temperatures occur in July, when rain is minimum.

The soil is loam with more clay in deeper horizons. In the orchard, four different pear varieties were cultivated, namely Williams, Abate, Kaiser, and Conference varieties, distributed in 17 rows with an inter-row distance of about 4 m, while the distance between plants on the row was about 1.5 m, depending on the variety; the soil, under the rows and between the rows, was grass-covered with periodic cutting.

2.1.3. Tomato Field

The third site included in the project is an area of about one hectare cultivated with industrial tomato "Pietra Rossa F1", inside the CREA research center in Montanaso Lombardo (LO). The site is characterized by loamy soils with an increase of clay with the depth, while climatic conditions are the same as the ones above described for the pear orchard site. Tomatoes were 0.3 m high at images' acquisition and cultivated in parallel rows of 0.5 m width with a distance between rows of about 1.5 m.

2.2. UAV Surveys and Photogrammetric Processing

Two UAV surveys were conducted on each study site, during the agricultural season of 2018. For each site, in the first survey, the Parrot Bebop 2 (Parrot S.A., Paris, France) was used to acquire

RGB images, and the Parrot Sequoia camera (Parrot S.A., Paris, France) was mounted on the Parrot Disco (Parrot S.A., Paris, France) to collect multispectral imagery during the second survey. The photogrammetric processing of all the surveys was performed with Pix4Dmapper Pro software (Pix4D S.A., Prilly, Switzerland), Version 4.1.1. The details of the surveys and their processing are described in the following sections.

2.2.1. Vineyard

The RGB survey of the vineyard took place on 28 June 2018, while the multispectral survey, six days later on 4 July. The flight height of the Parrot Bebop 2 used in the RGB survey was set to 40 m Above Ground Level (AGL), while the Parrot Disco flew at 60 m (AGL) during the multispectral survey. The same overlaps among images were fixed for the two surveys: longitudinal overlapping equal to 80% and transversal equal to 70%. An amount of 130 and 560 images were collected during the RGB and multispectral survey, respectively. According to the sensors' characteristics (i.e., Parrot Bebop 2 fisheye camera and Sequoia camera), the final Ground Sample Distance (GSD) of the acquired images was about 0.1 m for both cases. As suggested in [20], the coordinates of nine Ground Control Points (GCPs) were measured with the Global Navigation Satellite System (GNSS) receiver Leica Viva GS14 (Leica Geosystems, Heerbrugg, Switzerland) in Network Real-Time Kinematic (NRTK) mode, to ensure the georeferencing of the photogrammetric products with high accuracy. According to the results of [20], eight GCPs were distributed all around the perimeter of the vineyard, while the ninth target was placed in the middle of the field (Figure 2). In all the study sites, some of the GCPs were adopted as Check Points (CPs) during the photogrammetric processing, to assess process accuracy.

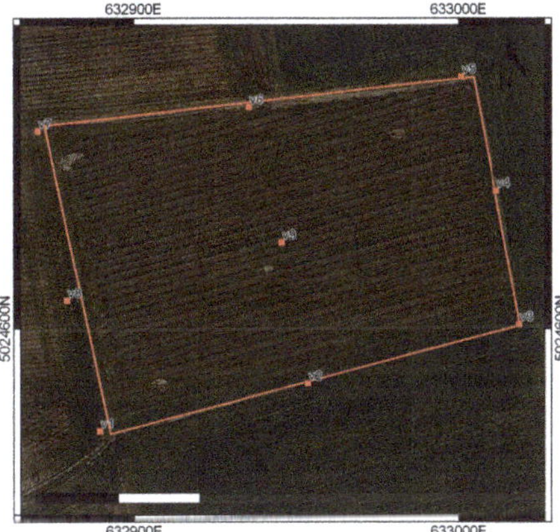

Figure 2. Ground Control Points' (GCPs) distribution for the surveys on the vineyard. Map data: ©Google Satellite.

At the end of two independent photogrammetric processing, performed with Pix4Dmapper Pro software, a Digital Surface Model (DSM), an RGB orthophoto, and a four band multispectral orthophoto were produced, with GSD of around 0.1 m, to exploit for the crop row detection. The DSM generated from the multispectral survey had lower quality than the RGB one; therefore, it was not considered in further analysis. Table 1 summarizes the residuals on GCPs after photogrammetric processing, while in Figure 3, DSM and orthophotos of the vineyard are shown.

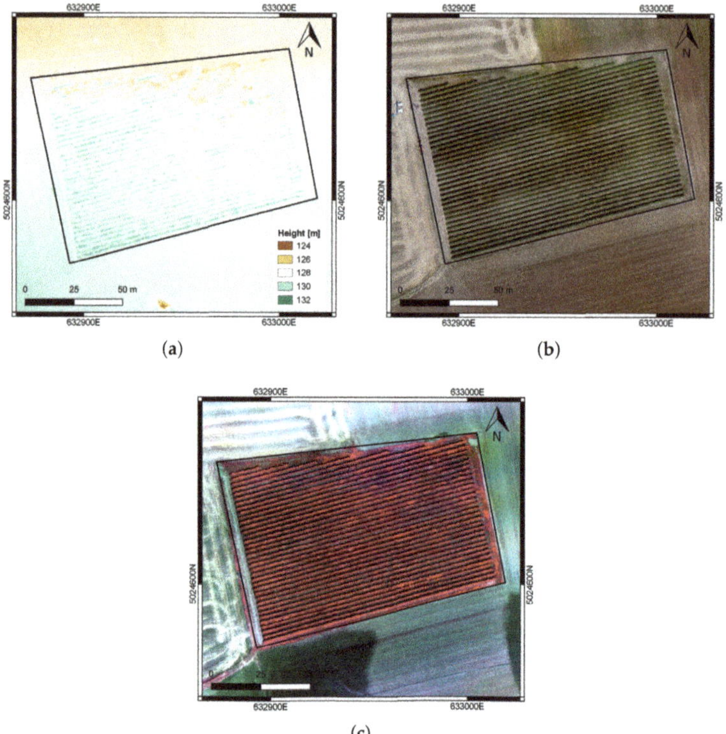

(**a**) (**b**)

(**c**)

Figure 3. Vineyard site: DSM (**a**) and orthophoto (**b**) produced through photogrammetric processing of the RGB dataset; false color orthophoto (**c**), generated from the multispectral dataset.

Table 1. Vineyard site: residuals on the GCPs after bundle block adjustment.

Label	Easting (m)	Northing (m)	Height (m)
v1	−0.029	−0.008	0.088
v2	−0.007	−0.012	−0.012
v3	−0.025	−0.030	−0.063
v4	0.008	0.025	0.062
v5	0.015	0.005	0.000
v6	0.022	−0.013	0.024
v7	0.030	−0.013	0.027
v8	0.020	−0.015	−0.086
v9	0.012	0.008	0.012
RMSE	**0.021**	**0.016**	**0.052**

2.2.2. Pear Orchard

In the pear orchard site, the RGB and multispectral surveys were conducted on 26 June and 2 July, respectively. The characteristics of the flights were the same as the surveys performed on the vineyard site: longitudinal overlap among images equal to 80%, transversal overlap equal to 70%, flight height set at 40 m and 60 m for the multirotor UAV and the fixed-wings UAV, respectively, thus ensuring a GSD of the images of about 0.1 m. 140 images were acquired during the RGB survey, while 540 during the multispectral one. During the UAV flights, seven targets were placed on the terrain to be used as GCPs, well distributed all around the orchard, as shown in Figure 4.

Figure 4. Ground Control Points' (GCPs) distribution for the surveys on the pear orchard. Map data: ©Google Satellite.

As the case of the vineyard, the DSM and the orthophotos were produced in Pix4Dmapper Pro with a spatial resolution of 0.1 m (Figure 5). The residuals on the GCPs computed after the photogrammetric workflow are reported in Table 2.

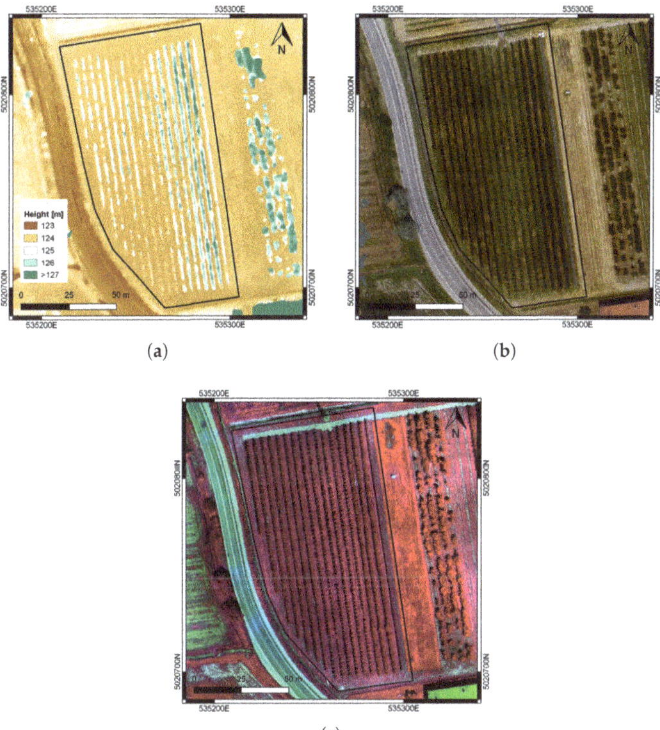

Figure 5. Pear orchard site: DSM (**a**) and orthophoto (**b**) produced through photogrammetric processing of the RGB dataset; false color orthophoto (**c**), generated from the multispectral dataset.

Table 2. Pear orchard site: residuals on the GCPs after bundle block adjustment.

Label	Easting (m)	Northing (m)	Height (m)
p1	−0.050	−0.017	0.007
p2	−0.032	−0.004	−0.017
p3	0.018	−0.060	0.021
p4	0.028	−0.089	0.113
p5	0.003	0.023	−0.290
p6	0.100	0.053	0.018
p7	−0.085	0.003	0.054
RMSE	**0.056**	**0.047**	**0.119**

2.2.3. Tomato Field

Given the proximity of the two sites, the tomato field was surveyed on the same days as the pear orchard, with the same equipment and flight characteristics. During the RGB survey, 96 images were acquired, as well as 388 images came from the multispectral flight. From these datasets, one DSM and two orthophotos, having a GSD equal to 0.1 m, were generated after bundle block adjustment in Pix4Dmapper Pro. The various photogrammetric products were georeferenced by means of six GCPs, whose center coordinates were measured with GNSS-NRTK on the dates of the surveys. The distribution of the GCPs on the tomato site and their computed residuals are shown in Figure 6 and Table 3, respectively. The photogrammetric products are reported in Figure 7.

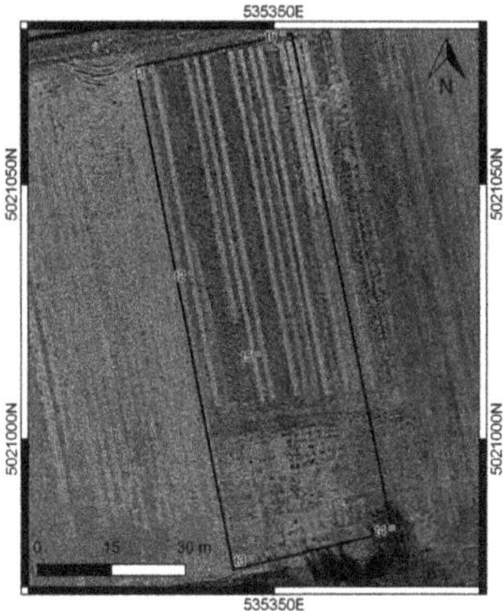

Figure 6. Ground Control Points' (GCPs) distribution for the surveys on the tomato site. Map data: ©Google Satellite.

Table 3. Tomato site: residuals on the GCPs after bundle block adjustment.

Label	Easting (m)	Northing (m)	Height (m)
t1	−0.098	0.071	−0.101
t2	0.095	0.068	0.046
t3	0.028	0.71	0.077
t4	−0.026	−0.127	−0.054
t5	−0.019	−0.154	0.149
t6	−0.053	−0.038	0.114
RMSE	**0.062**	**0.096**	**0.097**

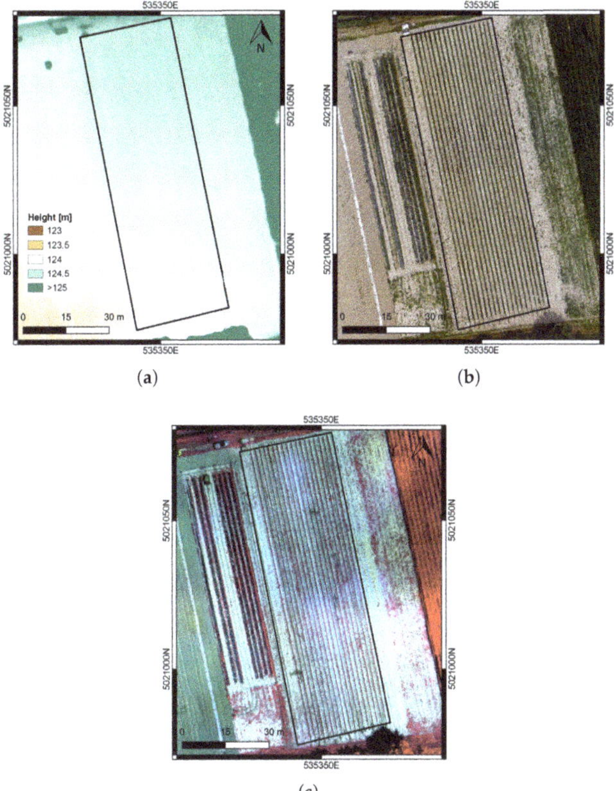

Figure 7. Tomato site: DSM (**a**) and orthophoto (**b**) produced through photogrammetric processing of RGB dataset; false color orthophoto (**c**), generated from the multispectral dataset.

3. Crop Row Detection Methods

Differentiating between crop canopy and background can be very challenging. Moreover, the different crop types considered in this study led to the need for many methods to extract crop rows, each one more suitable for a specific crop type. In the following sections, five different detection methods were proposed; some of them were taken from the existing literature (i.e., classification algorithms and Bayesian segmentation), while two methods, labeled as thresholding algorithms, were developed ad hoc for the purposes of the project.

In order to achieve the best possible crop mask to be used to identify crop rows on orthophotos, Vegetation Indices (VIs) were chosen as the inputs of the detection methods, as already proposed by many authors [15,16,21]. Considering vegetation, most of those indices take into account Red (R) and NIR reflectance bands (ρ_λ): the greater is the difference between ρ_R and ρ_{NIR}, the greater is the amount of green and healthy vegetation in that particular pixel. Among all the possible VIs, only those composed of spectral bands that sensors involved in the surveys could provide were used in this study (Table 4).

Table 4. Vegetation Indices (VIs) used in this study.

Index	Name	Formula	References
NDVI	Normalized Difference Vegetation Index	$\frac{NIR-Red}{NIR+Red}$	[22]
SR	Simple Ratio	$\frac{NIR}{Red}$	[23]
SAVI	Soil-Adjusted Vegetation Index	$\frac{NIR-Red}{NIR+Red+L}(1+L)$	[24]
ARVI	Atmospherically Resistant Vegetation Index	$\frac{NIR-RB}{NIR+RB}$ where: $RB = Red - \gamma(Blue - Red)$	[25]
ExG	Excess Green	$2(Green) - (Red + Blue)$	[26]
G%	Normalized Green Channel Brightness	$\frac{Green}{Red+Green+Blue}$	[27]

To exploit the proposed methods fully, also the DSM and RGB orthophoto were individually used as inputs. This ensured that the methods were still operational, even in the cases where only imagery resulting from UAVs supporting only RGB sensors was available, as for the Parrot Bebop 2.

The assessment of the final results was performed by computing error matrices and classification accuracies (Overall Accuracy (OA), User's Accuracy (UA), and Producer's Accuracy (PA)), on some validation samples manually identified on orthophotos, through visual inspection. In particular, the quality of the crop detection was defined according to the value of PA of the class crop canopy: the greater is the PA, the lower is the probability to omit crop pixels; therefore, to underestimate the detected crop rows.

3.1. Thresholding Algorithms

Two algorithms (specifically local maxima extraction and threshold selection) were developed ad hoc for the purpose of the project, by starting from [28], who proposed a method for crop rows' extraction by using as input the 3D point cloud. Both methods were generated in MATLAB R2017b [29] and are based on the concept that high pixel values generally correspond to crop rows. They are illustrated in the following sections.

3.1.1. Local Maxima Extraction

This method aims at generating a binary raster, where non-null values refer to the presence of the crop canopy. First, the input raster (e.g., VI map or DSM) is divided into square cells (macro-cells), then inside each macro-cell, the crop pixels are identified as those corresponding to a percentage of pixels with the highest values. It is a semi-automatic algorithm, where the user has to define the dimensions of the cell and the percentage value.

This method is sensitive to the user's choices: in particular, the macro-cell dimension should be selected in order to include both crop and ground pixels, thus being close to the distance between rows or slightly larger. Macro-cell size should be neither too small nor too big with respect to the distance between rows. If it is too small, a wrong selection of pixels is performed whatever the chosen percentage is. When the cell includes only crop pixels, whatever percentage not equal to 100% causes an underestimation of crop pixels; on the other side, when the cell overlaps only ground pixels, they would be selected as crop pixels, thus producing an over-estimation in the crop mask. The overestimation arises also when the dimension of the macro-cell is too big, because the probability of selecting pixels belonging to the ground increased with the chosen cell size.

3.1.2. Threshold Selection

This method produces a binary crop mask, by selecting as crop pixels all pixels with values higher than a reference value. The challenging part of this algorithm is the definition of this reference value. When both DSM and Digital Terrain Model (DTM) of the area are available, the Canopy Height Model (CHM) could be derived as difference between DSM and DTM, and zero could be considered as the reference value, while in other cases (i.e., VIs as input, no availability of an accurate DTM), it should be determined as described below.

Starting from the input raster, create a smoothed raster with a moving window average filter. Subtract the smoothed raster to the input raster, and define on the differences a threshold to be considered as the reference value, by visual checking of the results with an empirical trial and error approach. Pixels with values greater than the threshold are retained as crop. Even in this algorithm, the user intervention is twofold, choosing the dimensions of the moving window and the value of the threshold, and could cause problems of under-/over-estimation of crop canopy pixels.

3.2. *Classification Algorithms*

Two well-known classification algorithms were exploited in this study, K-means clustering and the Minimum Distance to the Mean (MDM) classifier, to be representative of both unsupervised and supervised classification algorithms. Both methods were applied in QGIS (Version 3.4) [30], to allow users not familiar with programming languages to run the algorithms thanks to a dedicated user-friendly GUI.

3.2.1. K-Means Clustering

This is a well-known algorithm for hard unsupervised thematic classification [31]. The clustering made by K-means is based on the minimization of the objective function $f(\Omega)$, defined as the Euclidean distance of samples of a cluster from the respective centroid.

The number of classes (K) are known a priori. Once K is defined, the method consists of three iterative steps. In the first step, for each class K_i, a centroid is automatically chosen. The rest of the data are assigned to K clusters based on the minimum distance criterion. The Euclidean distances of each sample from the centroids are computed, and in the second step, the sample is assigned to the cluster for which the computed distance is minimum. In the last step, centroids are re-calculated, and all the samples are re-assigned. This step is iterated until the clustering converges to a stable solution, namely when centroids of clusters do not change meaningfully.

The final configuration is stable and does not depend on the initial position of centroids arbitrarily selected. The initial configuration only influences the number of iterations necessary to reach the convergence.

3.2.2. Minimum Distance to Mean Classifier

This method finds the mean values of all the training sets and classifies all the image pixels according to the class mean they are closest. The process is performed for all image pixels, one at a time. Bounds are determined using statistics derived from the training sets, and the distance used is the Euclidean one.

As for all supervised algorithms, a crucial phase of the MDM classifier is the selection of the training samples. They are used to compute class spectral signatures, therefore must be representative of all the classes. In this study, training samples were defined by visual inspections of the UAV images and grouped in two macro-classes: crop canopy and background, which included weeds, soil, and shadow pixels.

3.3. Bayesian Segmentation

This method relies on the Bayesian approach, where any uncertainty can be considered random variables that are fully described by probability distributions [32]. Given the vector of data y and the vector of parameter x, the conditional distribution of parameters is described by the Bayes theorem [33]:

$$P(x|y) = \frac{P(x,y)}{P(y)} = \frac{P(y|x)P(x)}{P(y)} \tag{1}$$

where:
$P(x|y)$ is called the posterior probability and describes the new level of knowledge of the unknown parameters x given the observed data y.
$P(y)$ is a normalization constant used to impose that the sum of $P(y|x)$ for all possible x is equal to one.
$P(x)$, instead, represents the prior probability distribution. It describes the knowledge of the unknown parameters x without the contribution of the observed data.
$P(y|x)$ is defined as the likelihood and is a function of x. It describes the way in which the a priori knowledge is modified by data and depends on the noise distribution.

The terms in Equation (1) can be adapted to match the purpose of this study, the detection of crop rows: the posterior probability is the probability of a pixel to be part of the class crop canopy or background, and the prior probability is defined starting from the mean and standard deviation values, a priori assigned to each class, while the likelihood is described by a Gaussian distribution, in which the parameters are the mean and standard deviation of the two classes:

$$P(y_i|x_i) = \frac{1}{\sigma_{x_i}\sqrt{2\pi}} exp\left(-\frac{(y_i - \mu_{x_i})^2}{2\sigma_{x_i}^2}\right) \tag{2}$$

The final goal of Bayesian approach can be identified in finding the optimal parameters x that maximize the posterior probability distribution $P(x|y)$. This is called the Maximum A Posteriori (MAP) estimate [34], and it is defined as:

$$x^{MAP} = \arg\max_x P(x|y) \tag{3}$$

In crop row detection, it consists of assigning a unique class to each pixel of the image, depending on the posterior probabilities estimated for each pixel. In order to obtain outputs less affected by pixel noise, smoothing filters or image adjustment can be applied on the input raster.

4. Results

Considering all the detection methods, their parameters, and all the possible input rasters, the number of crop masks potentially available is very high. For the vineyard site, 191 outputs were tested and 166 and 104 for the pear orchard and tomato site, respectively. For the sake of brevity, only the best masks, representing each detection method, are reported and compared. An exhaustive analysis of all the tests performed was described in [35].

4.1. Vineyard

For the vineyard site, the parameters of each detection method that resulted in the best results are reported in Table 5.

Table 5. Vineyard site: parameters for the best results of each detection method. MDM, Minimum Distance to the Mean.

Method	Input	User's Choices
Local Maxima Extraction	G%	cell size: 5 m percentage: 30%
Threshold Selection	DSM	cell size: 3 m threshold: 0.3
K-means Clustering	RGB orthophoto	classes: 6
MDM Classifier	RGB orthophoto	classes: 2
Bayesian Segmentation	ExG, Gaussian filter (σ = 3)	Background: μ = 0.2, σ = 0.2 Crop canopy: μ = 0.7, σ = 0.25

For the computation of the error matrices, 104 polygons (N pixels = 42,732) were defined for the class crop canopy and 97 polygons (N pixels = 64,309) for the class background. In Table 6, the accuracies of the five selected best results are summarized, and the detail of each crop mask is shown in Figure 8.

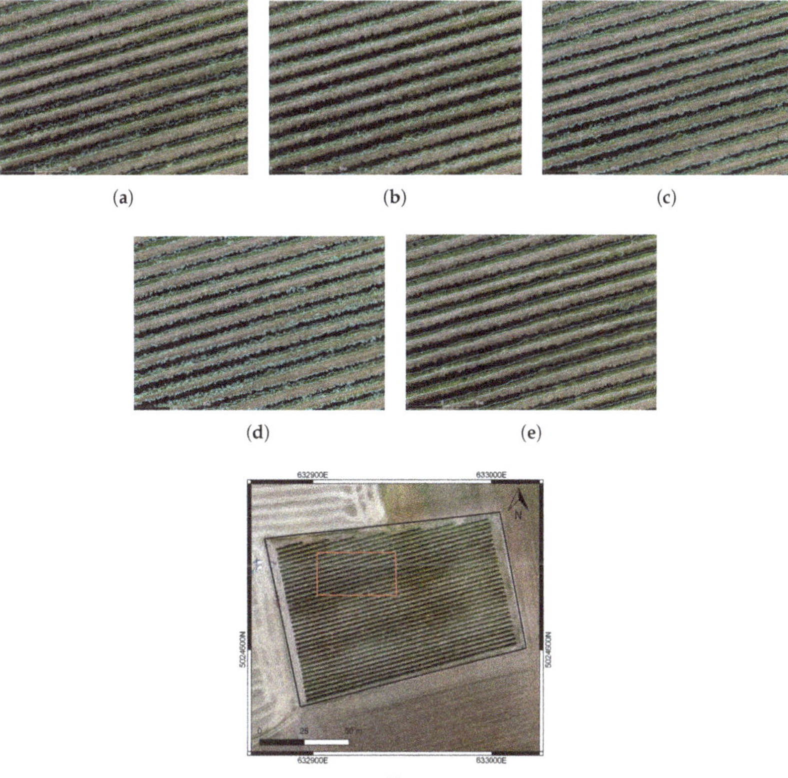

Figure 8. Vineyard site: crop row detection results for (**a**) local maxima extraction, (**b**) threshold selection, K-means clustering (**c**), MDM classifier (**d**), and the Bayesian segmentation (**e**). Figures refer to the area included in the red box in (**f**).

Table 6. Vineyard site: assessment of the best results of each detection method.

Method	OA	PA Crop Canopy	UA Crop Canopy
Local Maxima Extraction	0.94	0.95	0.91
Threshold Selection	0.76	0.41	0.99
K-means Clustering	0.82	0.73	0.80
MDM Classifier	0.87	0.84	0.83
Bayesian Segmentation	0.96	0.97	0.94

4.2. Pear Orchard

The parameters for the best results of each detection method for the pear orchard are summarized in Table 7.

Table 7. Pear orchard site: parameters for the best results of each detection method.

Method	Input	User's Choices
Local Maxima Extraction	DSM	cell size: 4 m percentage: 40%
Threshold Selection	DSM	cell size: 4 m threshold: 0
K-means Clustering	RGB orthophoto	classes: 5
MDM Classifier	RGB orthophoto	classes: 2
Bayesian Segmentation	NDVI, Gaussian filter ($\sigma = 3$)	Background: μ=0.8, σ=0.04 Crop canopy: μ=0.93, σ=0.04

The error matrices were computed starting from a validation set composed by 37 polygons (N pixels = 44,889) for the class crop canopy and 34 polygons (N pixels = 62,378) for the class background. The five selected best results and their respective accuracies are presented in Figure 9 and in Table 8.

Table 8. Pear orchard site: assessment of the best results of each detection method.

Method	OA	PA Crop Canopy	UA Crop Canopy
Local Maxima Extraction	0.92	0.88	0.93
Threshold Selection	0.95	0.97	0.92
K-means Clustering	0.95	0.90	0.99
MDM Classifier	0.87	0.68	0.99
Bayesian Segmentation	0.94	0.91	0.95

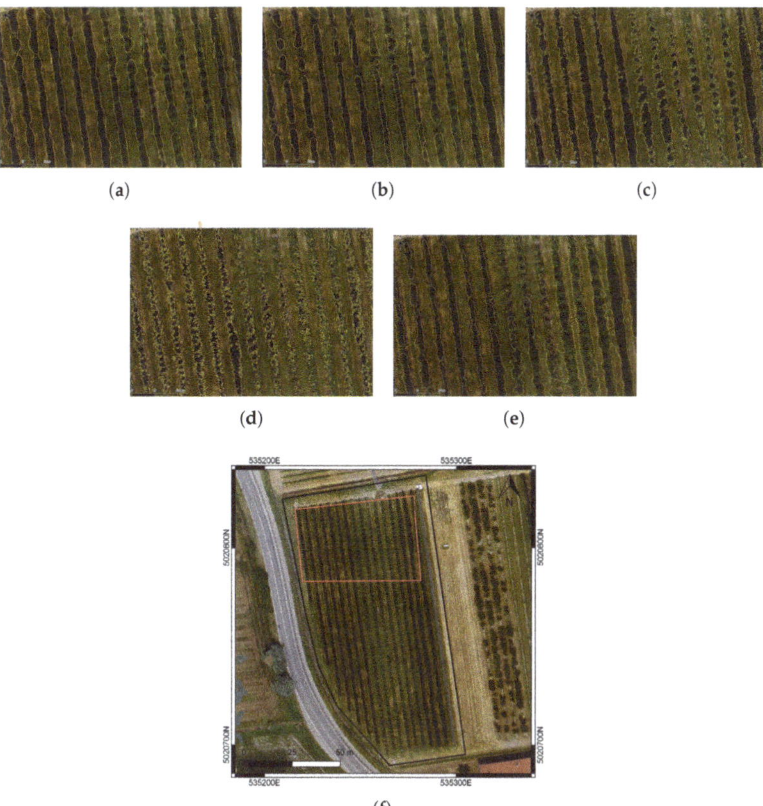

Figure 9. Pear orchard site: crop row detection results for (a) local maxima extraction, (b) threshold selection, K-means clustering (c), MDM classifier (d), and Bayesian segmentation (e). Figures refer to the area included in the red box in (f).

4.3. Tomato Field

Table 9 reports the parameters chosen for each detection method, which gave the best crop mask outputs.

Table 9. Tomato field site: parameters for the best results of each detection method.

Method	Input	User's Choices
Local Maxima Extraction	G%	cell size: 3 m percentage: 30%
Threshold Selection	DSM	cell size: 4 m threshold: 0
K-means Clustering	SAVI + NDVI	classes: 5
MDM Classifier	SAVI + NDVI	classes: 2
Bayesian Segmentation	ExG, Histogram adjustment	Background: $\mu = 0.05$, $\sigma = 0.15$ Crop canopy: $\mu = 0.65$, $\sigma = 0.35$

The assessment of the results for the tomato site was performed on 52 polygons (N pixels = 81,290) for the class crop canopy and 42 polygons (N pixels = 155,296) for the class background. The accuracy values are presented in Table 10, while Figure 10 shows the crop detection for the five selected methods.

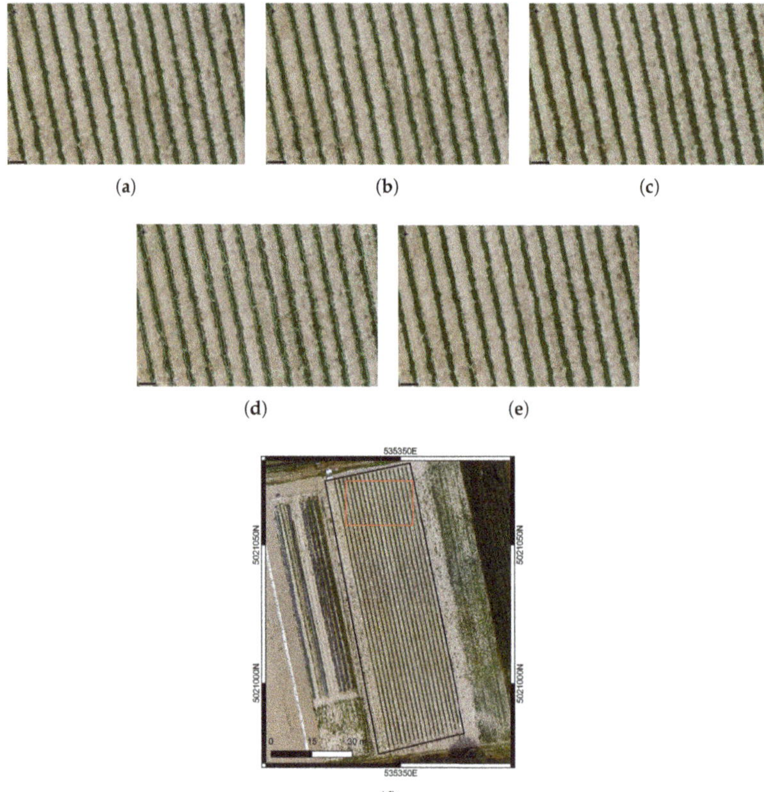

Figure 10. Tomato field site: crop row detection results for (**a**) local maxima extraction, (**b**) threshold selection, K-means clustering (**c**), MDM classifier (**d**), and Bayesian segmentation (**e**). Figures refer to the area included in the red box in (**f**).

Table 10. Tomato field site: assessment of the best results of each detection method.

Method	OA	PA Crop Canopy	UA Crop Canopy
Local Maxima Extraction	0.98	0.94	0.98
Threshold Selection	0.97	0.87	0.99
K-means Clustering	0.93	0.93	0.79
MDM Classifier	0.90	0.60	0.92
Bayesian Segmentation	0.98	0.91	0.98

5. Discussion

As general findings, it can be stated that all the methods tested in this study performed well for crop row detection, with OA close or even greater than 0.9. The vineyard site seemed to be the most challenging (OA values lower than 0.9 for some methods), due to the concurrent presence of weed, bare soil, and shadow in the inter-row distance, while the high contrast between bare soil and crop canopy facilitated the crop detection in the tomato site (OA values always higher than 0.9).

Despite the high accuracy values achieved, the proposed algorithms did not require any particular computational resources, and the calculation time is reasonable with mass-market hardware, even if it varied according to the method. Local Maxima Extraction (LME) and Bayesian Segmentation (BS) overall returned the best outputs in terms of accuracies values, but with different performances in terms of time cost and parameter setting. The first method was faster, and choosing a cell size comparable with the rows' distance, or slightly larger, and a percentage of maxima between 30% and 40% could produce high quality results in all cases. The latter required a high level of a priori knowledge, and parameters had to be ad hoc fine-tuned with a time-consuming trial and error approach. The Threshold Selection (TS) algorithm needs to be run with an accurate DSM, and the definition of the reference value is crucial and sometimes can fail, as demonstrated by the low accuracy values registered in the vineyard site (OA = 0.76), especially on fields characterized by a relevant slope. On hilly fields and non-flat areas, the use of a real CHM is necessary and cannot be bypassed by the creation of a smoothed raster, on which the reference value of terrain height is identified. Classification algorithms are easy to run, especially in the QGIS implementation, and widely used in remote sensing, but cannot reach, in all cases, the same level of accuracy as the other methods. In addition, these algorithms require considerable human intervention, either in the labeling phase, as the case of K-means clustering, or in the delineation of the training samples, as for starting the MDM Classifier.

Regarding input rasters, DSM, RGB orthophotos, and VIs obtained as a combination of RGB bands are the most adopted in the selected methods. Only the cases of Bayesian segmentation on the pear orchard site and classification algorithms for tomato site require NDVI and NDVI jointly with SAVI to obtain the best results. Therefore, NIR information does not give any particular additional value in crop row detection, and RGB sensors can perform accurate canopy extraction, as already demonstrated by other authors [15,28], saving the time and cost of the UAV surveys and processing.

According to crop characteristics, specific considerations can be stressed for each single crop. In the case of a vineyard, it is important to maintain the continuity of the crop row, when detecting the crop canopy. This characteristics was enhanced in the Bayesian segmentation, as shown in Figure 8e, also thanks to the Gaussian filter applied to the input raster before launching the algorithm. The continuity of the vine rows was also guaranteed by using the local maxima extraction algorithm as the detection method (Figure 8a), apart from some rare and sparse pixels. Indeed, the two aforementioned methods registered the highest accuracy values, in particular PA values: 0.97 and 0.95 for BS and LME, respectively, considerably greater than the PA values of the other detection methods. The major issue of detecting vine rows is the presence of shadows, weeds, and bare soil in the inter-row distance. Our results demonstrated that the shadows made the classification methods practically unusable on the vineyard. In Figure 8c,d, it is clearly visible how the pixels at the edges of the shadow areas were detected as crop pixels. Classification algorithms are unable to separate the vine canopy from its shadow on the terrain, resulting in an overestimation of the crop rows (UA values around 0.8).

In the orchard, pear trees are planted quite distant from one another (in our study site, around 1.5 m); therefore, a good detection has to identify single plants rather than rows. In these terms, classification algorithms return the best results, as visible in Figure 9c,d and confirmed by the highest values of UA practically equal to one (Table 8). On the other hand, these detection methods also returned the most noisy outputs and underestimated the presence of pear trees in the orchard, in particular the MDM classifier with a PA equal to 0.68. The height of the trees favored their extraction from the background, also in presence of weeds in the inter-row distance. To exploit these characteristics fully, it is advisable to use the DSM as the input of whatever detection method; in particular, the threshold selection algorithm gave the best outcomes in the pear orchard site, with PA value for the class crop canopy of 0.97.

As already mentioned, the tomato field site had the highest accuracy values for the crop detection, thanks to the regular alternation of bare soil and vegetation canopy. The OA values for all the tested methods were higher than 0.9; the LME algorithm, BS, and K-means also returned PA values above 0.9, while the PA of the TS method was slightly lower than 0.9 due to the use of the DSM as the input for starting the algorithm. At the time of the survey, the plants had a height of 0.3 m, equivalent to a few image GSDs; therefore, the errors in the photogrammetric processing (column height in Table 3) in this specific case affected the results of the canopy detection.

Precision viticulture is already widespread in the world, and recent articles have demonstrated the added value that remote sensing from UAV platforms can give to this sector [36]. Hence, numerous studies can be found in the literature dealing with vine canopy extraction [15,28,37,38]. The results presented in this work had accuracy values similar to those available in the literature. To the best of authors' knowledge, very few studies have already been published related to the detection of pear trees in orchards or tomato canopy, thus hampering the availability of reference values to compare the outputs. The assessment of the reliability of the illustrated results was based on similar case studies present in the literature. The detection of pear plants was performed with accuracy values slightly lower than the results obtained by [39], in two orchards in China, but close to the outcomes of the chestnut tree extraction described in [16]. The high values found for the tomato site were in agreement with the results presented in [17], about the estimation of crop emergence in potatoes.

The potential utility of this study in precision agriculture is high. The methods herein described allowed deriving from UAV imagery vegetation properties specifically related to the characteristics of the crop under investigation. In particular, in the irrigation management context, it must be taken into account that usually, irrigation for row-crops is provided through localized pressurized systems, wetting only the areas near the crop. The soil between rows is usually grass-covered, so analyzing crop maps (for example NDVI or thermal maps) of the field without masking the soil between rows could lead to errors when evaluating the water status or vigor of the crop under study. In Figures 11–13, NDVI maps for the three analyzed study sites are shown: on the left, the original maps, while on the right, the canopy maps generated after the extraction of crop rows. This information could be used in precision agriculture applications for mapping vegetation stress status or to optimize on-farm irrigation management. As an example, in [40], the use of a crop row detection method to delineate Site-Specific Management Zones (SSMZs) maps on a vineyard was described.

Moreover, it is crucial to stress that in the case of row-crops, the importance of using UAV imagery with respect to satellite imagery becomes fundamental. In satellite images, in fact, pixels have larger dimensions, and the operation of extracting crop rows is not possible. This leads to the presence of mixed pixels, for which Vegetation Indices' maps give information about both row and inter-row vegetation, therefore not very usable for agronomic inputs' management [41,42].

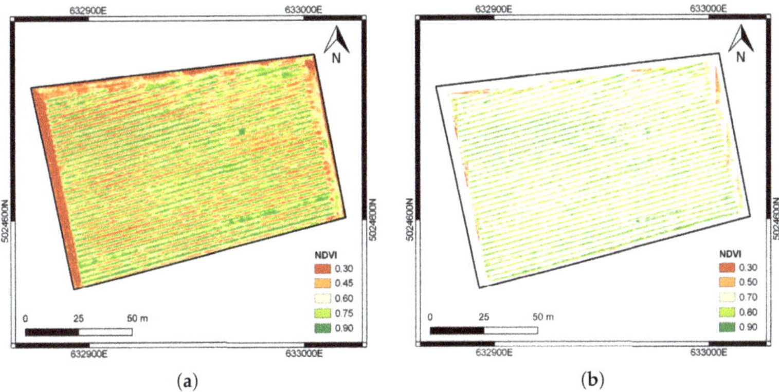

Figure 11. Vineyard site: NDVI map before (**a**) and after (**b**) the crop rows' extraction.

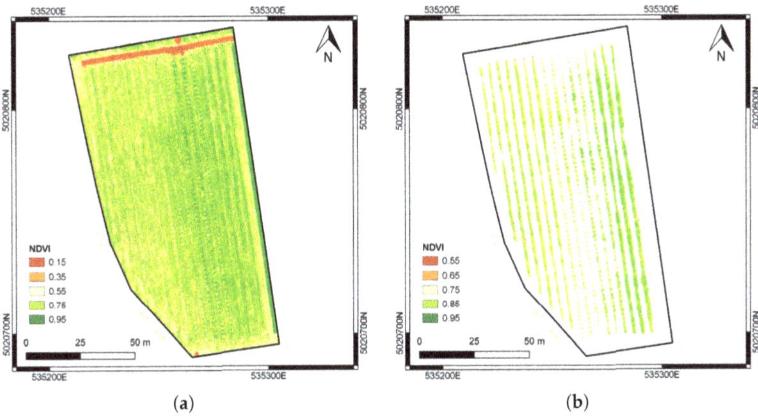

Figure 12. Pear orchard site: NDVI map before (**a**) and after (**b**) the crop rows' extraction.

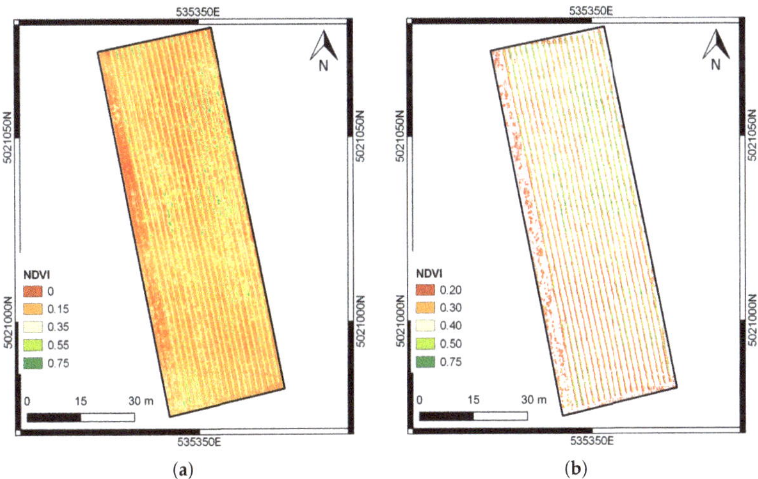

Figure 13. Tomato field site: NDVI map before (**a**) and after (**b**) the crop rows' extraction.

6. Conclusions

This study demonstrates the feasibility to perform crop row detection from high-resolution UAV imagery, for different crop types, including vineyards, orchards, and horticultural crops. DSM, RGB, or multispectral orthophotos can be used as input for the detection methods; in particular, the DSM performs better with crop characterized by high heights (i.e., grapevine and pear), even in the presence of inter-row weed, but it should be avoided to detect horticultural crops (i.e., tomato). Commercial RGB sensors give high accuracy values for crop row detection; therefore, for this purpose, it is not necessary to perform surveys mounting more expensive multispectral cameras, if no additional infrared information is required. Furthermore, in the presence of shadows produced by the crop canopy on the terrain, indices based on the NIR band and classification algorithms could lead to an overestimation of the crop rows.

Although all applied methods need some level of human intervention, among all, the local maxima extraction algorithm, developed ad hoc within this study, allows reaching the best compromise in terms of time-cost, automation, and quality of results. Bayesian segmentation applied on VIs performs better than the other methods in the presence of bare soils, but it depends on a priori information.

Author Contributions: Conceptualization, G.R., A.F., B.O., and G.S.; methodology, G.R. and G.S.; software, G.R. and A.M.; validation, G.R.; data curation, G.R., A.M.; writing, original draft preparation, G.R. and A.M.; writing, review and editing, G.R., A.M., B.O., A.F., and G.S.; visualization, G.R. and A.M.; supervision, B.O., A.F., and G.S.; project administration, B.O., A.F., and G.S.; funding acquisition, B.O., A.F., and G.S. All authors read and agreed to the published version of the manuscript.

Funding: We wish to thank Regione Lombardia for funding the NUTRIPRECISO project (EU-RDP 2017), in the context of which this research was developed.

Conflicts of Interest: The authors declare no conflict of interest. The funders had no role in the design of the study; in the collection, analyses, or interpretation of data; in the writing of the manuscript; nor in the decision to publish the results.

References

1. Pachauri, R.K.; Allen, M.R.; Barros, V.R.; Broome, J.; Cramer, W.; Christ, R.; Church, J.A.; Clarke, L.; Dahe, Q.; Dasgupta, P.; et al. *Climate Change 2014: Synthesis Report. Contribution of Working Groups I, II and III to the Fifth Assessment Report of the Intergovernmental Panel on Climate Change*; IPCC: Geneva, Switzerland, 2014.
2. Costa, J.; Vaz, M.; Escalona, J.; Egipto, R.; Lopes, C.; Medrano, H.; Chaves, M. Modern viticulture in southern Europe: Vulnerabilities and strategies for adaptation to water scarcity. *Agric. Water Manag.* **2016**, *164*, 5–18. [CrossRef]
3. UN-Water. *2018 UN World Water Development Report, Nature-Based Solutions for Water*; UNESCO, Paris, France, 2018.
4. Castellarin, S.D.; Bucchetti, B.; Falginella, L.; Peterlunger, E. Influenza del deficit idrico sulla qualità delle uve: Aspetti fisiologici e molecolari. *Italus Hortus* **2011**, *18*, 63–79.
5. Monaghan, J.M.; Daccache, A.; Vickers, L.H.; Hess, T.M.; Weatherhead, E.K.; Grove, I.G.; Knox, J.W. More 'crop per drop': Constraints and opportunities for precision irrigation in European agriculture. *J. Sci. Food Agric.* **2013**, *93*, 977–980. [CrossRef]
6. Sanchez, L.; Sams, B.; Alsina, M.; Hinds, N.; Klein, L.; Dokoozlian, N. Improving vineyard water use efficiency and yield with variable rate irrigation in California. *Adv. Anim. Biosci.* **2017**, *8*, 574–577. [CrossRef]
7. McClymont, L.; Goodwin, I.; Whitfield, D.; O'Connell, M. Effects of within-block canopy cover variability on water use efficiency of grapevines in the Sunraysia irrigation region, Australia. *Agric. Water Manag.* **2019**, *211*, 10–15. [CrossRef]
8. Matese, A.; Di Gennaro, S.F. Technology in precision viticulture: A state of the art review. *Int. J. Wine Res.* **2015**, *7*, 69–81. [CrossRef]
9. Pádua, L.; Vanko, J.; Hruška, J.; Adão, T.; Sousa, J.J.; Peres, E.; Morais, R. UAS, sensors, and data processing in agroforestry: A review towards practical applications. *Int. J. Remote Sens.* **2017**, *38*, 2349–2391. [CrossRef]

10. Nebiker, S.; Annen, A.; Scherrer, M.; Oesch, D. A light-weight multispectral sensor for micro UAV: Opportunities for very high resolution airborne remote sensing. *Int. Arch. Photogramm. Remote Sens. Spat. Inf. Sci.* **2008**, *37*, 1193–1199.
11. Tsouros, D.C.; Bibi, S.; Sarigiannidis, P.G. A Review on UAV-Based Applications for Precision Agriculture. *Information* **2019**, *10*, 349. [CrossRef]
12. Corti, M.; Cavalli, D.; Cabassi, G.; Vigoni, A.; Degano, L.; Gallina, P.M. Application of a low-cost camera on a UAV to estimate maize nitrogen-related variables. *Precis. Agric.* **2019**, *20*, 675–696. [CrossRef]
13. Noh, H.; Zhang, Q. Shadow effect on multi-spectral image for detection of nitrogen deficiency in corn. *Comput. Electron. Agric.* **2012**, *83*, 52–57. [CrossRef]
14. Pádua, L.; Marques, P.; Hruška, J.; Adão, T.; Bessa, J.; Sousa, A.; Peres, E.; Morais, R.; Sousa, J.J. Vineyard properties extraction combining UAS-based RGB imagery with elevation data. *Int. J. Remote Sens.* **2018**, *39*, 5377–5401. [CrossRef]
15. Poblete-Echeverría, C.; Olmedo, G.; Ingram, B.; Bardeen, M. Detection and segmentation of vine canopy in ultra-high spatial resolution RGB imagery obtained from unmanned aerial vehicle (UAV): A case study in a commercial vineyard. *Remote Sens.* **2017**, *9*, 268. [CrossRef]
16. Marques, P.; Pádua, L.; Adão, T.; Hruška, J.; Peres, E.; Sousa, A.; Sousa, J.J. UAV-based automatic detection and monitoring of chestnut trees. *Remote Sens.* **2019**, *11*, 855. [CrossRef]
17. Li, B.; Xu, X.; Han, J.; Zhang, L.; Bian, C.; Jin, L.; Liu, J. The estimation of crop emergence in potatoes by UAV RGB imagery. *Plant Methods* **2019**, *15*, 15. [CrossRef] [PubMed]
18. Pádua, L.; Adão, T.; Sousa, A.; Peres, E.; Sousa, J.J. Individual Grapevine Analysis in a Multi-Temporal Context Using UAV-Based Multi-Sensor Imagery. *Remote Sens.* **2020**, *12*, 139. [CrossRef]
19. Ortuani, B.; Facchi, A.; Mayer, A.; Bianchi, D.; Bianchi, A.; Brancadoro, L. Assessing the Effectiveness of Variable-Rate Drip Irrigation on Water Use Efficiency in a Vineyard in Northern Italy. *Water* **2019**, *11*, 1964. [CrossRef]
20. Ronchetti, G.; Pagliari, D.; Sona, G. DTM generation through UAV survey with a fisheye camera on a vineyard. *Int. Arch. Photogramm. Remote Sens. Spat. Inf. Sci.* **2018**, *XLII-2*, 983–989. doi:10.5194/isprs-archives-XLII-2-983-2018. [CrossRef]
21. Pádua, L.; Marques, P.; Hruška, J.; Adão, T.; Peres, E.; Morais, R.; Sousa, J. Multi-Temporal Vineyard Monitoring through UAV-Based RGB Imagery. *Remote Sens.* **2018**, *10*, 1907. [CrossRef]
22. Rouse J., Jr.; Haas, R.; Schell, J.; Deering, D. Monitoring Vegetation Systems in the Great Plains with ERTS. *NASA Spec. Publ.* **1974**, *351*, 309.
23. Birth, G.S.; McVey, G.R. Measuring the color of growing turf with a reflectance spectrophotometer 1. *Agron. J.* **1968**, *60*, 640–643. [CrossRef]
24. Huete, A.R. A soil-adjusted vegetation index (SAVI). *Remote Sens. Environ.* **1988**, *25*, 295–309. [CrossRef]
25. Kaufman, Y.J.; Tanre, D. Atmospherically resistant vegetation index (ARVI) for EOS-MODIS. *IEEE Trans. Geosci. Remote Sens.* **1992**, *30*, 261–270. [CrossRef]
26. Woebbecke, D.M.; Meyer, G.E.; Von Bargen, K.; Mortensen, D. Color indices for weed identification under various soil, residue, and lighting conditions. *Trans. ASAE* **1995**, *38*, 259–269. [CrossRef]
27. Richardson, A.D.; Jenkins, J.P.; Braswell, B.H.; Hollinger, D.Y.; Ollinger, S.V.; Smith, M.L. Use of digital webcam images to track spring green-up in a deciduous broadleaf forest. *Oecologia* **2007**, *152*, 323–334. [CrossRef]
28. Weiss, M.; Baret, F. Using 3D point clouds derived from UAV RGB imagery to describe vineyard 3D macro-structure. *Remote Sens.* **2017**, *9*, 111. [CrossRef]
29. MATLAB. *Version 9.3 (R2017b)*; The MathWorks Inc.: Natick, MA, USA, 2017. Available online: https://www.mathworks.com/downloads/ (accessed on 14 December 2018).
30. QGIS Development Team. *QGIS Geographic Information System*. Open Source Geospatial Foundation, 2009. Available online: http://qgis.osgeo.org (accessed on 2 January 2020).
31. MacQueen, J. Some methods for classification and analysis of multivariate observations. In Proceedings of the Fifth Berkeley Symposium on Mathematical Statistics and Probability, Oakland, CA, USA, 1 January 1967; Volume 1, pp. 281–297.
32. Ross, S.M. *Probabilità e statistica per l'ingegneria e le scienze*; Apogeo Editore, Milano, Italy, 2003.

33. Bayes, T. LII. An essay towards solving a problem in the doctrine of chances. By the late Rev. Mr. Bayes, FRS communicated by Mr. Price, in a letter to John Canton, AMFR S. *Philos. Trans. R. Soc. Lond.* **1763**, *53*, 370–418. [CrossRef]
34. Geman, S.; Geman, D. Stochastic relaxation, Gibbs distributions, and the Bayesian restoration of images. *IEEE Trans. Pattern Anal. Mach. Intell.* **1984**, 721–741. [CrossRef]
35. Marino, A.; Marotta, F. Crop rows detection through UAV images. In *Master Degree Final Dissertation, Politecnico di Milano*; 2019. Available online: https://www.politesi.polimi.it/ (accessed on 28 November 2019).
36. BorgognoMondino, E.; Gajetti, M. Preliminary considerations about costs and potential market of remote sensing from UAV in the Italian viticulture context. *Eur. J. Remote Sens.* **2017**, *50*, 310–319. [CrossRef]
37. Cinat, P.; Di Gennaro, S.F.; Berton, A.; Matese, A. Comparison of Unsupervised Algorithms for Vineyard Canopy Segmentation from UAV Multispectral Images. *Remote Sens.* **2019**, *11*, 1023. [CrossRef]
38. de Castro, A.; Jiménez-Brenes, F.; Torres-Sánchez, J.; Peña, J.; Borra-Serrano, I.; López-Granados, F. 3-D characterization of vineyards using a novel UAV imagery-based OBIA procedure for precision viticulture applications. *Remote Sens.* **2018**, *10*, 584. [CrossRef]
39. Dong, X.; Zhang, Z.; Yu, R.; Tian, Q.; Zhu, X. Extraction of Information about Individual Trees from High-Spatial-Resolution UAV-Acquired Images of an Orchard. *Remote Sens.* **2020**, *12*, 133. [CrossRef]
40. Ortuani, B.; Sona, G.; Ronchetti, G.; Mayer, A.; Facchi, A. Integrating Geophysical and Multispectral Data to Delineate Homogeneous Management Zones within a Vineyard in Northern Italy. *Sensors* **2019**, *19*, 3974. [CrossRef] [PubMed]
41. Di Gennaro, S.F.; Dainelli, R.; Palliotti, A.; Toscano, P.; Matese, A. Sentinel-2 Validation for Spatial Variability Assessment in Overhead Trellis System Viticulture Versus UAV and Agronomic Data. *Remote Sens.* **2019**, *11*, 2573. [CrossRef]
42. Khaliq, A.; Comba, L.; Biglia, A.; Ricauda Aimonino, D.; Chiaberge, M.; Gay, P. Comparison of satellite and UAV-based multispectral imagery for vineyard variability assessment. *Remote Sens.* **2019**, *11*, 436. [CrossRef]

Article

Evaluation of Rapeseed Winter Crop Damage Using UAV-Based Multispectral Imagery

Łukasz Jełowicki [1], Konrad Sosnowicz [2], Wojciech Ostrowski [3], Katarzyna Osińska-Skotak [3,*] and Krzysztof Bakuła [3]

1 OPEGIEKA Sp. z o.o., 82-300 Elbląg, Poland; lukasz.jelowicki@opegieka.pl
2 Skysnap Sp. z o.o., 02-001 Warsaw, Poland; konrad.sosnowicz@skysnap.pl
3 Department of Photogrammetry, Remote Sensing and Spatial Information Systems, Faculty of Geodesy and Cartography, Warsaw University of Technology, 00-661 Warsaw, Poland; wojciech.ostrowski@pw.edu.pl (W.O.); krzysztof.bakula@pw.edu.pl (K.B.)
* Correspondence: katarzyna.osinska-skotak@pw.edu.pl

Received: 30 June 2020; Accepted: 12 August 2020; Published: 13 August 2020

Abstract: This research is related to the exploitation of multispectral imagery from an unmanned aerial vehicle (UAV) in the assessment of damage to rapeseed after winter. Such damage is one of a few cases for which reimbursement may be claimed in agricultural insurance. Since direct measurements are difficult in such a case, mainly because of large, unreachable areas, it is therefore important to be able to use remote sensing in the assessment of the plant surface affected by frost damage. In this experiment, UAV images were taken using a Sequoia multispectral camera that collected data in four spectral bands: green, red, red-edge, and near-infrared. Data were acquired from three altitudes above the ground, which resulted in different ground sampling distances. Within several tests, various vegetation indices, calculated based on four spectral bands, were used in the experiment (normalized difference vegetation index (NDVI), normalized difference vegetation index—red edge (NDVI_RE), optimized soil adjusted vegetation index (OSAVI), optimized soil adjusted vegetation index—red edge (OSAVI_RE), soil adjusted vegetation index (SAVI), soil adjusted vegetation index—red edge (SAVI_RE)). As a result, selected vegetation indices were provided to classify the areas which qualified for reimbursement due to frost damage. The negative influence of visible technical roads was proved and eliminated using OBIA (object-based image analysis) to select and remove roads from classified images selected for classification. Detection of damaged areas was performed using three different approaches, one object-based and two pixel-based. Different ground sampling distances and different vegetation indices were tested within the experiment, which demonstrated the possibility of using the modern low-altitude photogrammetry of a UAV platform with a multispectral sensor in applications related to agriculture. Within the tests performed, it was shown that detection using UAV-based multispectral data can be a successful alternative for direct measurements in a field to estimate the area of winterkill damage. The best results were achieved in the study of damage detection using OSAVI and NDVI and images with ground sampling distance (GSD) = 10 cm, with an overall classification accuracy of 95% and a F1-score value of 0.87. Other results of approaches with different flight settings and vegetation indices were also promising.

Keywords: damage detection; winter crop; UAV; multispectral imagery; vegetation indices; agricultural insurance; technical roads

1. Introduction

The European Union is the global leader in rapeseed production, and Poland is one of the largest rapeseed producers and processors in Europe. In 2019, it was second only to France [1]. In 2010–2019, the area of rapeseed sown in Poland ranged from 720,000 to 950,000 hectares, representing about 8–10% of the total crops, and about 95–96% of the oilseed acreage [2].

Due to the climate changes observed in recent years (i.e., snow-free winters), yield losses caused by poor wintering of plants are becoming more frequent. Low temperatures are the most dangerous for crops, combined with no snow or a thin layer of snow cover. Rapid warming causing thaws, followed by frosts, are also not conducive to wintering plants. The result of such winter weather is freezing winter crops, i.e., damage to the crops. If there is no snow cover, winter rapeseed may freeze at −10 °C [3]. Annual losses in rapeseed yields in 2015–2019 were estimated to range from 10% to 20% [2]. To assess damages for agricultural insurance, an assessment of its occurrence in the field is performed. These are imprecise measurements made based on a sample selected in the field whereby an expert determines the number of plants per square meter and takes this as a representation of one hectare of the affected field. The results of these measurements depend on the expert's experience. In the case of rapeseed crops, compensation is awarded when there are less than 15 healthy plants per square meter and the damage covers at least 10% of the field area. The damage threshold values (min. 10% and max. 50% of the parcel area) are particularly important because, according to the scope of insurance, they affect the final decision on the qualification of loss in crops (partial or total compensation). Since the methods of assessing damages used by insurance companies are not very accurate and are time consuming, it appears that the use of remote sensing techniques may be an excellent alternative.

Currently, research conducted in the area of remote sensing applications with the use of unmanned aerial vehicles (UAV) is often related to aspects of vegetation monitoring, particularly relating to forest and agricultural areas. UAV-based remote sensing in agriculture is particularly related to pest and disease detection, development of crops during their growth cycle, assessment of biomass, and water stress in plants [4]. To date, these phenomena have been assessed through satellite and airborne photogrammetry [5–8]. The breakthrough solutions in terms of image acquisition techniques are those using UAVs [9–11]. There is a need to receive continuous spatial information with a high level of accuracy and actuality, while retaining low operating costs. Drones can provide an alternative, providing high resolution images with a short revisit time, even every couple of hours, contrary to optical satellite systems or aerial images which have a much lower time resolution.

To date, analyses connected with object geometry, such as an assessment of parcel areas, landslide monitoring, and cropland and forest inventory, have been widely applied in many countries [12–14]). However, applications of sensors with more than three spectral bands (e.g., multispectral or hyperspectral cameras) mounted on unmanned aerial vehicles are not as commonly used. Research on implementation of these cameras and integration of various remote sensing techniques is a key area of interest for many groups of scientists [15–18].

Remote sensing techniques for vegetation mapping play an important role in precise agriculture [9,19] and phenotyping research of crops [20–22]. Remote monitoring of arable lands using spectral libraries (obtained from field measurements, satellite imagery, or aerial and close-range images) could be helpful in the analysis of plant growth [19,22], assessment of the size of harvest [23] and crop fertilizing needs [24,25], pest control [24], and the extraction of dead plants. For many years, the Remote Sensing Centre at the Institute of Geodesy and Cartography in Poland has carried out monitoring of farmland with the use of satellite imagery to forecast harvests [26]. The accuracy of the prepared models for the prediction of crop volumes using NOAA AVHRR data is estimated at 90–95%. A high correlation between a harvest's size and its spectral characteristics based on field measurements [27] is a clear indicator of the potential of remote sensing data for such research. This translates into numerous areas of scientific research, involving spectral vegetation indices in the assessment of the physical condition of flora [5,19,27,28].

The use of spectral indices aims to extract essential information about vegetation condition, and the indices should indicate a proper correlation between their values and biophysical parameters of plant cover, such as biomass or leaf area index (LAI) [5,19,27,29]. The most popular spectral indices in this area are those that use big reflectance differences between the near infrared region and the red band. This kind of index includes the normalized difference vegetation index (NDVI) and simple ratio (SR). Based on the research carried out to date, it is observed that there is a link between the above-mentioned parameters and the size of the biomass and LAI [5]. The outcome of scientific studies has confirmed that a high accuracy of the performed estimations (85%) [18,19] was found, and that there are strong and statistically significant relationships between NDVI (obtained from UAV) and crop biophysical variables (for LAI, R^2 was 0.95; for ground truth percent canopy cover, R^2 was 0.93). The obtained strong relationships suggest that UAV multispectral data can be used for estimation of LAI and percent canopy cover. Wei et al. (2017) [30] concluded in their research that derivative spectral indices formulated using optimized narrow wavebands were most effective in quantifying the changes in pigment and water content of leaves subjected to freezing injury.

Frost is one of the environmental stresses of plants. The plant response to frost can be a decrease in chlorophyll, inhibition of photosynthesis, altered leaf angle, and plant freezing [31]. The influence of frost on biochemical and biophysical changes, as well as spectral properties of various plant species, have been investigated [30,32–36]. It was found that freezing causes changes in the spatial differentiation of chlorophyll content on the lamina surface [32] and the structural changes of mesophyll cells [30]. In turn, these changes affect the spectral reflection of various plant parts, especially in blue-green, green, red, and near infrared radiation [30,33–36]. However, as the analyzed studies show, these changes depend on the species studied and the degree of freezing.

In order to analyze the effects of crop freezing in winter, it is essential that the influence of the spectral reflectance of soil should be included. For practical implementation, it is common to use the soil-adjusted vegetation index (SAVI). Because of the difficulty in choosing the optimal value of the coefficient describing soil brightness and the low sensitivity to a small amount of chlorophyll, the modified chlorophyll absorption ratio index (MCARI) was introduced. This parameter highlights an amount of chlorophyll absorption in the range of about 670 nm in relation to spectral characteristics centered at 550 and 700 nm [6]. With the use of band normalization in the spectral range of red edge and red (R700, R670), it is possible to minimize the influence of soil and extract information about vegetation without green pigments. Another method of implementation is to use the transformed chlorophyll absorption reflectance index (TCARI)/optimized soil adjusted vegetation index (OSAVI). The use of normalization enables the detection of biomass variability even at low LAI values [6]. Furthermore, Hunt's research (2008) [28] confirms the effectiveness of the green normalized difference vegetation index (GNDVI, [37]), which differentiates biomass content within the area of winter wheat crops. Moreover, studies carried out by many scientists provides evidence that GNDVI correlates significantly with the nutrient ingredients (mainly nitrogen compounds) [28,38]. Among the methods used, it is worth mentioning the fusion of chosen spectral indices. As an example, the use of NDVI, the modified soil-adjusted vegetation index (MSAVI)/the transformed soil-adjusted vegetation index (TSAVI) (in the case where spectral bands in the shortwave infrared (SWIR) range are available) and GNDVI enables the assessment of the spatial variability of cropland, including its impact on agricultural yields. The opportunities from using remote sensing data outlined above appear to be a reasonable solution to crop insurance, including estimating the winterkill losses. Multispectral sensors have been significantly miniaturized in recent years, and their current price allows for the commercialization of the presented method without significant financial outlay. In contrast to satellite data, the quality of which is determined by cloud cover range, the use of UAVs in time-series analysis is not as strongly related to weather conditions. This fact is essential for regions of moderate climate zones because winter crop inventories are conducted in the late autumn which is usually very foggy. Furthermore, an application of the appropriate methodology for data processing and analysis enables farmers and insurance providers to accurately estimate winterkill losses and the amounts of compensation paid by insurers. One of the most important steps in this process is damage detection at the beginning of the vegetation period.

The purpose of this study was to: (1) evaluate the possibility of using UAVs for estimating losses in rapeseed crops caused by poor wintering, (2) determine the impact of data acquisition parameters (mainly the altitude influencing the ground sampling distance (GSD)), and (3) present the methodology of data processing (proper vegetation index) to obtain the best results compared to the traditional method of field inventory.

2. Materials and Methods

The experiment was carried out to analyze the possibility of using UAV-based multispectral imagery in the detection of damaged rapeseed after winter and the estimation of the damage area. The experiment was carried out in cooperation with an insurance company. Figure 1 shows the scheme of the experiment performed. Data were registered during three test flights on different flight heights in order to evaluate the influence of image spatial resolution. Each time, remote sensing data were acquired in four spectral bands, namely green, red, red edge, and near infrared, which were used to calculate tested vegetation indices. RGB images were acquired and combined with high-resolution spectral data, which were later used for manual preparation of reference data. Training samples were obtained during the field campaign.

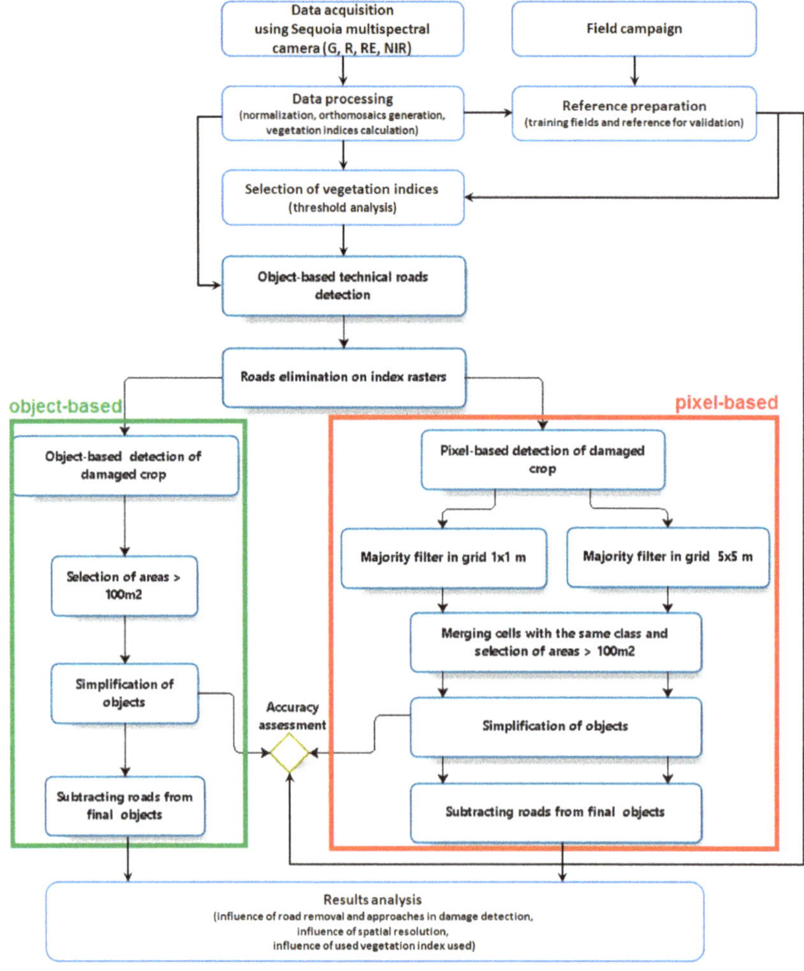

Figure 1. The scheme of the experiment performed.

In the methodology, there were a few crucial elements which are described in this section. Firstly, selected indices were tested to evaluate their usefulness in further analysis. The important part of the experiment was to delineate technical roads and replace those areas with a mean value from neighborhoods to exclude the effect of machinery influence (technical roads) on estimating damage. Another issue in the experiment was to define the influence of spatial resolution of the multispectral data on the obtained results.

2.1. Test Area and Data Used

For the planned tests, a 34 ha parcel with a rapeseed field was selected. It is situated in Olecko County in Northern Poland (Figure 2). This region is well known for its low temperatures and harsh climate throughout the year, hence, it was chosen for this experiment on the detection of winter crop damage. The winter of 2016/2017 in this region was characterized by high variability of weather conditions, with alternate periods of frost and thaw. During the period of severe frosts (from −12 to −27 °C) there was a thin layer of snow of several centimeters. This caused winter damage, which was confirmed during a field inspection.

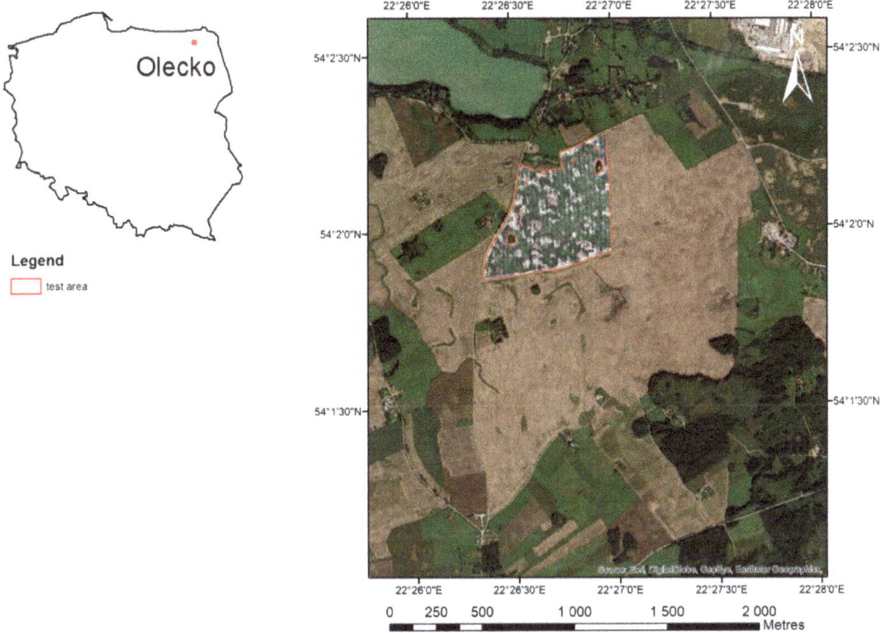

Figure 2. Test area, rapeseed field in Olecko, Poland.

The testing data were acquired with the use of an eBee UAV equipped with a Sequoia multispectral camera. This allowed for the acquisition of data in four different spectral bands: green (530–570 nm), red (640–680 nm), red edge (730–740 nm), and near infrared (770–810 nm). Three flights were performed on 20 April 2017, at three different altitudes of 50, 100, and 200 m above the ground. This resulted in images with three different spatial resolutions—ground sampling distances (GSD) of 5, 10, and 20 cm, respectively. This allowed us to assess the impact of spatial resolution on the imagery in the detection of a crop's health when undamaged by frost. The overlapping of successive images along a flight strip (endlap) was 80%, and the overlap between flight strips (sidelap) was 70%. Flights were carried out in midday hours (12:00 p.m.–2:00 p.m.), with a cloudless or slightly cloudy sky. Wind speed was 1–2 m/s, air temperature was 5–6 °C, and humidity was 32–37%. The visibility was 35.5–43.8 km.

The orientation of images was performed in Pix4DMapper Pro software (Pix4D SA, Switzerland). This software allows the user to select the appropriate calculation template. For multispectral data calculation, the AG (agriculture) multispectral template is preferred. The program automatically recognizes the camera, retrieves its parameters from the database, groups the images into the appropriate spectral bands, and enables them to be calibrated. A useful feature of the Sequoia camera is that it can measure the amount of irradiance coming from the sun for each image and for each band with the sunshine sensor. This enables the user to take irradiance into account and normalize the value of the images amongst themselves. In the orientation process, self-calibration with 15 ground control points (control and check points) measured with a GNSS receiver was also performed to give a final root mean square error (RMS) of aerial triangulation of 10 cm. The result of the images processed in Pix4DMapper Pro were reflectance orthomosaics.

Based on the orthomosaics generated from the collected images in four spectral bands, different vegetation indices were calculated, based on which further experiments were conducted. To perform the correct classification of healthy and damaged crops, 23 training fields were created using data from field inspection and verified on RGB images. The training fields contained information about the number of healthy plants over a surface of 1 square meter (Figure 3). This helped to describe the relationship between the vegetation indices values and the number of plants in a surface unit, which resulted in thresholds for image classification.

Figure 3. Examples of the training fields (1 m × 1 m).

2.2. Selection of Vegetation Indices

The most important process in determining post-winter damage is the correct detection of areas where the crop has not grown properly, or where the plants have died. Remote sensing as a multidisciplinary technique is a tool with many indices, which can help test the health of plants and solve the problem of the value of compensation for farmers for damaged crops. The Sequoia camera was the equipment used on the UAV in this study and allowed recording in four spectral bands: green, red, red edge, and near infrared. As a result of a literature review, a set of vegetation indicators was identified for further analysis. Consequently, after preliminary tests, the final selection was based on three basic spectral indices of vegetation: NDVI, SAVI, and OSAVI. In addition, an interesting solution in the calculation of these indices was to replace the red channel with the red edge, which was supported by the multispectral camera used in the experiment. Such modification was carried out and three further indices were examined: NDVI_RE, SAVI_RE, OSAVI_RE.

The NDVI (normalized difference vegetation index) is one of the most basic and commonly used indices [39]. It is a measure of photosynthetic activity and determines the condition of the plant and its development stage. Plants absorb most of the red light that hits it while reflecting much of the near infrared light. When vegetation is dead or stressed it reflects more red light and less near infrared light. Based on these properties, NDVI is calculated as the normalized difference between the red and near infrared bands from an image. The basic formula for calculating the index value is as follows:

$$\mathrm{NDVI} = \frac{\mathrm{NIR} - \mathrm{RED}}{\mathrm{NIR} + \mathrm{RED}} \tag{1}$$

where NIR is the near infrared band value and RED is the red band value.

The SAVI (soil adjusted vegetation index) is a modification of the NDVI index [40]. Due to changes in the numerator and denominator of the typical NDVI equation (soil brightness correction factor), the SAVI index further improves the final result for the influence of the soil brightness. For this reason, it is used in situations where a large part of the crop is not covered with lush vegetation. Therefore, the beginning of the growing season appears to be an excellent time for examining the condition of plants using this index. The formula for calculating SAVI is as follows:

$$\text{SAVI} = \frac{(1+\text{L}) * (\text{NIR} - \text{RED})}{\text{NIR} + \text{RED} + \text{L}} \tag{2}$$

where L denotes the coverage of the vegetation area. In most cases, and particularly for the case of intermediate vegetation canopy levels, optimal results are achieved using L = 0.5. For this experiment, a value 0.5 was used during the tests.

The OSAVI (optimized soil adjusted vegetation index) was developed by Rondeaux et al. (1996) [41] and was first presented in the work entitled "Optimized Soil Adjusted Vegetation Index". This index is a modification of the SAVI that has been optimized for agricultural monitoring. It is more sensitive to changes in plant condition than the original SAVI and does not need a priori knowledge of the soil type. A value of 0.16 as the soil adjustment coefficient was selected as the optimal value to minimize variation with soil background. The formula for calculating OSAVI is as follows:

$$\text{OSAVI} = \frac{\text{NIR} - \text{RED}}{\text{NIR} + \text{RED} + 0.16} \tag{3}$$

In addition to the above three spectral indices, their modifications were tested in which the red channel was replaced by red edge. By analyzing spectral reflectance properties for plants in red edge wavelength, we can observe quick changes in reflection and can thus expect to see more discreet differences between healthy and damaged plants. To test this premise, NDVI_RE, SAVI_RE, and OSAVI_RE indices were calculated. The application of such indices with a modified channel has been discussed previously [42,43].

The correctness of damage recognition by the vegetation index was examined using training fields measured during a field inventory with the participation of insurance specialists. Twenty-tree sites were surveyed with an area of approximately 1 square meter and the condition of the plants was determined in their range. These fields were then characterized by experts as to whether they would qualify for compensation or if the plants in the field area were healthy. A training field was also specified that could be treated as two classes on the border: damaged and healthy crops. The training fields were ranked according to the condition of the plants. After calculating the mean values of the tested indices for each training field, the values obtained were compared with the reference from the field inventory. The threshold value was established based on the training fields which, according to experts, were found not to meet the conditions for granting compensation. Based on the healthy training field, threshold values for each index were used to classify (or not) training fields for compensation. All training fields were verified with an expert in the field. Their correctness was ensured so the outlier fields were eliminated in defining the thresholds. The threshold value was calculated as the mean value from the lowest mean value of the index for an accepted healthy crop training field and the highest mean value of the index for an accepted damaged crop training field. A comparison of results for training fields with defined thresholds is shown in Figure 4. As can be seen, the training fields were characterized by differentiation of the values of the vegetation indices. The highest standard deviation was noted in the fields with healthy plants, and the lowest in damaged areas. The greatest variability of the index values was related to the occurrence of plants of different sizes and condition. Training field no. 9 recorded higher NDVI and SAVI values due to the presence of weeds; furthermore, in training field no. 11, there were 10 healthy, well-developed plants, which also resulted in higher NDVI values.

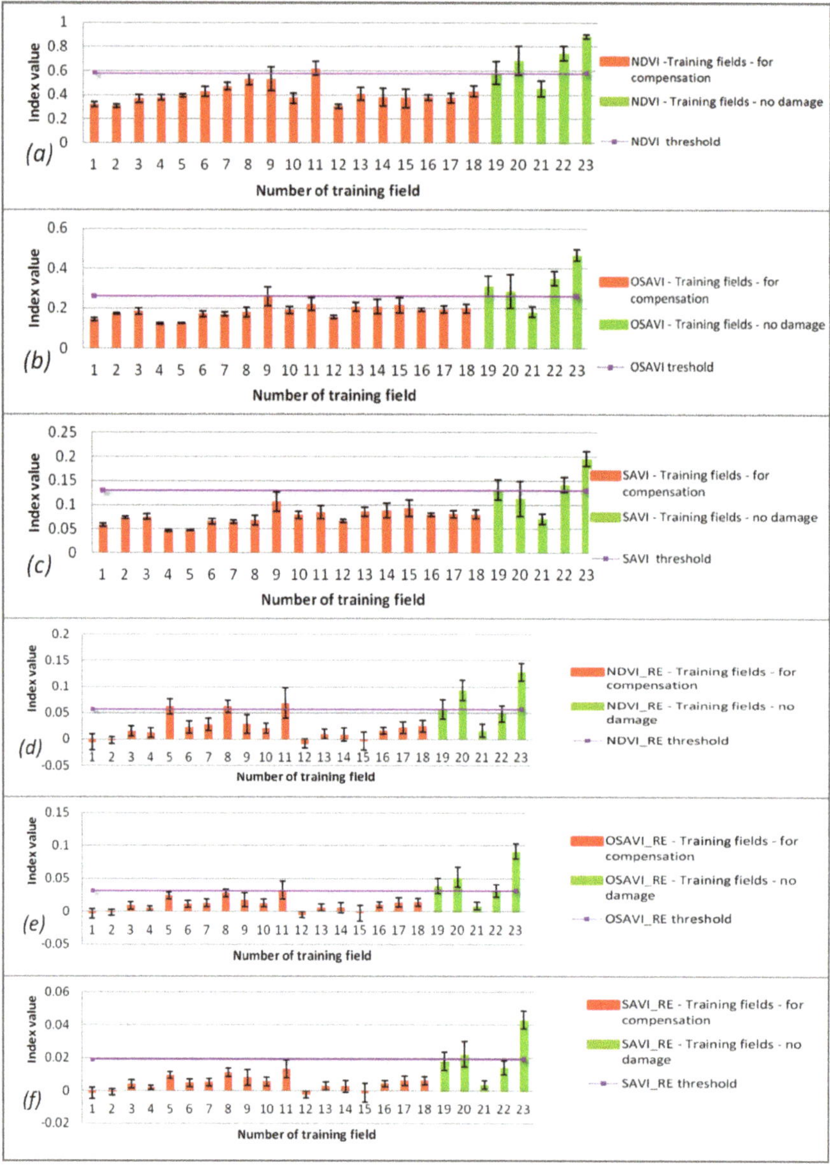

Figure 4. Interpretation of training fields with selected thresholds according to indices: Normalized Difference Vegetation Index (NDVI) (**a**), Optimized Soil Adjusted Vegetation Index (OSAVI) (**b**), Soil Adjusted Vegetation Index (SAVI) (**c**), Normalized Difference Vegetation Index—Red Edge (NDVI_RE) (**d**), Optimized Soil Adjusted Vegetation Index—Red Edge (OSAVI_RE) (**e**), and Soil Adjusted Vegetation Index—Red Edge (SAVI_RE) (**f**). The training fields are ranked according to the condition of plants. The chart shows the average value of index for each of the training fields with the standard deviation.

Based on the selected thresholds, referring to each index, Figure 5 illustrates the errors in training fields, verified by inspection of the field. The red areas represent the test fields that, according to specialists, should be classified as damaged areas. The green areas represent fields without visible frozen influence. The threshold shown by a horizontal line in Figure 4 determined the correctness of interpretation for each training field: red areas should be below the threshold and green above. If the situation is reversed, it means that tested index does not work properly. After summarizing the errors, the top 3 results were finally selected, i.e., NDVI, OSAVI, and OSAVI_RE, to carry out further experiments.

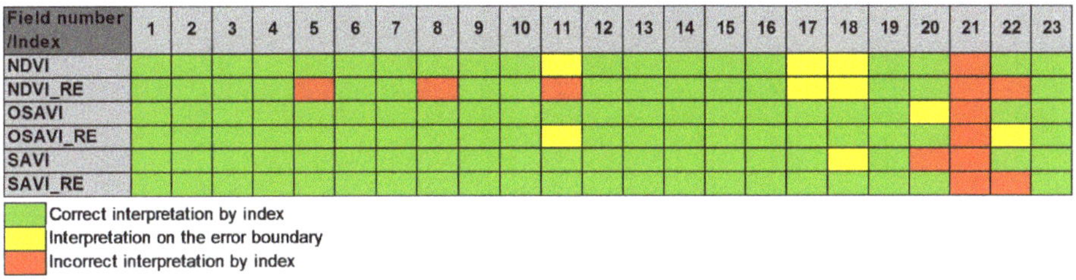

Field number /Index	1	2	3	4	5	6	7	8	9	10	11	12	13	14	15	16	17	18	19	20	21	22	23
NDVI	Green	Green	Green	Green	Green	Green	Green	Green	Green	Green	Yellow	Green	Green	Green	Green	Green	Yellow	Yellow	Green	Green	Red	Green	Green
NDVI_RE	Green	Green	Green	Green	Red	Green	Green	Red	Green	Green	Red	Green	Green	Green	Green	Green	Yellow	Yellow	Green	Green	Red	Red	Green
OSAVI	Green	Green	Green	Green	Green	Green	Green	Green	Green	Green	Green	Green	Green	Green	Green	Green	Green	Green	Green	Yellow	Red	Green	Green
OSAVI_RE	Green	Green	Green	Green	Green	Green	Green	Green	Green	Green	Yellow	Green	Green	Green	Green	Green	Green	Green	Green	Green	Red	Yellow	Green
SAVI	Green	Green	Green	Green	Green	Green	Green	Green	Green	Green	Green	Green	Green	Green	Green	Green	Green	Yellow	Green	Red	Red	Green	Green
SAVI_RE	Green	Green	Green	Green	Green	Green	Green	Green	Green	Green	Green	Green	Green	Green	Green	Green	Green	Green	Green	Green	Red	Red	Green

Green: Correct interpretation by index
Yellow: Interpretation on the error boundary
Red: Incorrect interpretation by index

Figure 5. Errors of the classification for training fields intended for compensation process.

In addition, considering the training fields, it is clear that the worst results were obtained for field no. 21. This was a unique area with healthy but very small plants (significant influence of the soil background). This suggests that the accepted threshold method may be sensitive to the stage in which the plants are grown, and results in slight over detection of damaged areas.

2.3. *Elimination of Road Influence*

Regardless of the selected vegetation index, the influence of technical roads is clearly visible on the classification results. The classification of technical roads, which are visible as traces of tires on a part of the field in poor condition, could have a big impact on the final results of detection, because this area should not be included in the calculation of the area of rapeseed damaged by frost.

The main insurer's condition applied to an area for reimbursement is that the area of a detected polygon must cover some minimum area (i.e., larger than 100 m^2). Therefore, many smaller objects are removed from the analysis. However, small isolated damaged areas are sometimes related to technical roads, which can result in detection of one bigger area, and are thus mistakenly detected as an area eligible for reimbursement. To prevent such situations, it was decided to work on rasters without the influence of roads on the reimbursement area.

For this purpose, road detection was conducted, based on compositions consisting of NDVI and green, red, and red edge channels. Object-based classification was performed in eCognition software for each dataset (data with spatial resolution equal to 5, 10, and 20 cm). In such a classification, information about spectral values, object shape, and extent were used. This resulted in three different road images, one for each spatial resolution of 5, 10, and 20 cm. Detected road images differed significantly from one another. The most precise and complete images came from images of 5 cm, and the worst from the lowest resolution (see Figure 6).

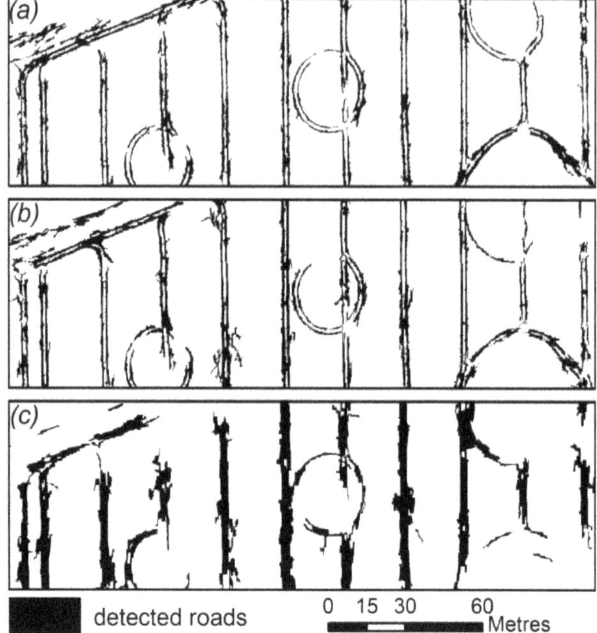

Figure 6. Detected roads on orthomosaics with ground sampling distance (GSD) of 5 cm (**a**), 10 cm (**b**), and 20 cm (**c**).

The biggest problems with areas of roads are presented in Figure 7, where there is no visible border between road and bare soil, which is why some tires traces were not detected. However, these places do not need to be detected because they do not cause the described problem of connecting polygons. Subsequently, a road mask was subtracted from the index rasters and the resulting holes were replaced with a mean value from corresponding cells (see Figure 7). Thus, the negative influence of technical roads on the final results of classification and detection of regions for reimbursement was reduced.

2.4. Approaches for Classification

The detection of damaged areas was performed using three different vegetation indices (NDVI, OSAVI, OSAVI_RE) in three different classification approaches, which were used to avoid the influence of technical roads on the result of winter damage crop estimation. The first method was a simple object-based classification to delineate the areas eligible for reimbursement. The main assumption behind this method was that it should appropriately suit borders of reference polygons which represent real situations and were vectorized without the generalization of polygons.

The second and third approach exploited pixel-based classification with thresholds. Index images were intersected with grids of cell sizes of 1 × 1 m and 5 × 5 m. For cells, a mean value of the index was assigned, and each cell was classified and merged with corresponding cells with the same class, creating homogenous regions. Thus, the resulting polygon geometry was simplified. By using those two approaches, we could also simulate a decrease in the resolution of input data.

In every approach, objects larger than 100 m^2 were finally recognized as detected damages. However, smaller regions lying within 5 m of other suspected damaged objects were also used, if the sum of their area was bigger than 100 m^2.

The possibility of proper detection of crops damaged by frost was examined using many variants. The influence of road elimination was examined, in addition to using a vegetation index for detection and spatial resolution of collected data. This resulted in 42 variants.

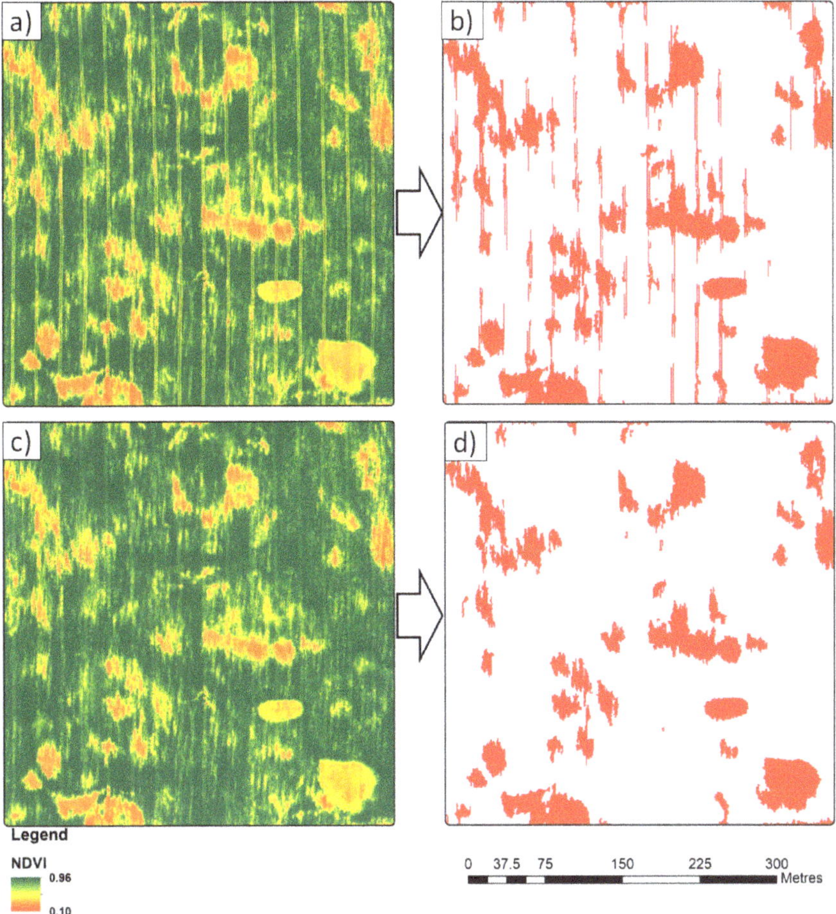

Figure 7. Presentation of impact of tires traces shown on NDVI images and the results of classification of winter damage for crops: (**a**) NDVI image, (**b**) NDVI image after elimination of influence of roads, (**c**,**d**) the results of classification of winter damages for crop based on the NDVI image, before and after elimination of influence of roads.

2.5. Accuracy Assessment

To estimate the accuracy of the final indications of which areas in each variant were eligible for reimbursement, a reference image was created by vectorization of damaged crops on an orthomosaic (see Figure 8).

The results were compared with reference areas by creating an error matrix which allowed for calculating such parameters as producer's accuracy (PA, completeness), user's accuracy (UA, correctness), F1-score, and overall accuracy (OA) [44,45]. The F1-score is the harmonic mean of precision and sensitivity and is usually used as an accuracy measure of a dichotomous model [45], so it is suitable for one-class delineation. In compensation assessment, the accurate area size of detected damage is much more important for the insurance company than its position, so another parameter was also used. This was the basic ratio of detected damaged areas to the reference area, and for the purpose of this study we named it "area index" (AI).

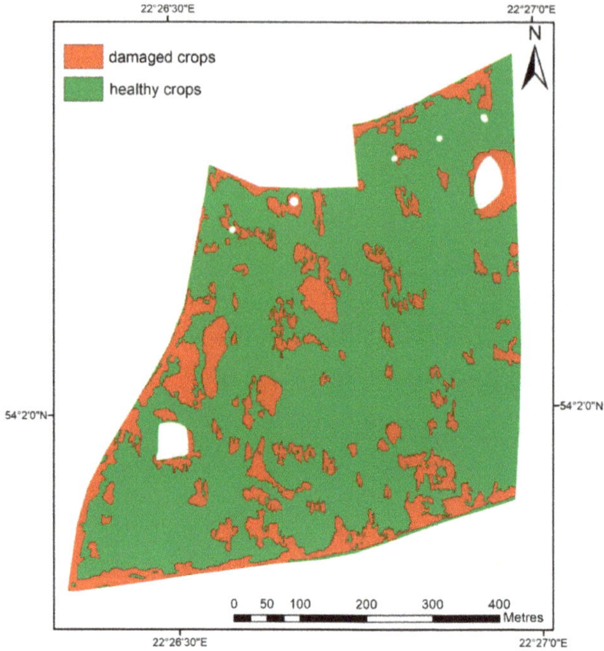

Figure 8. Reference data: crop field map manually vectorized.

3. Results

The results of all variants for damaged rapeseed detection are presented in Table A1 in Appendix A. Figure 9 illustrates the comparison of the overall accuracy, the F1-score, and AI for all variants of the experiment. The visual effect of the classification for selected variants is presented in Figure 10.

Most variants of classification allow detection of winterkill with OA above 90% (Figure 9a). UA and PA ranged between 70–80% and 90–100%, respectively (Table A1 in Appendix A). This means that the algorithm detects almost every polygon from our reference set but also mistakenly indicates other areas (overestimation). The poorer result in UA may indicate some mistakes in the reference set, which was difficult to vectorize objectively in a field, and because the algorithm may overestimate the detection of crop losses, which is confirmed by the area of damage often exceeding 110% to 130% (Figure 9c). However, it should be noted that the overestimation relates to transition areas, which are inherently difficult to clearly separate (see Figure 11). Generally, the best results were achieved using OSAVI and NDVI indices (the highest UA, OA, and F1-score values). The highest accuracy (measured by F1-score and OA) with the smallest overestimation of the area of losses in crops (AI) was obtained for a spatial resolution of 10 cm. The worst results (low OA and F1, and high AI) were obtained for OSAVI_RE, using images with GSD = 5 cm. The area for compensation was overestimated by 29–31%, which may be caused, for example, by inaccuracies in the georeference of an image with a spatial resolution of 5 cm.

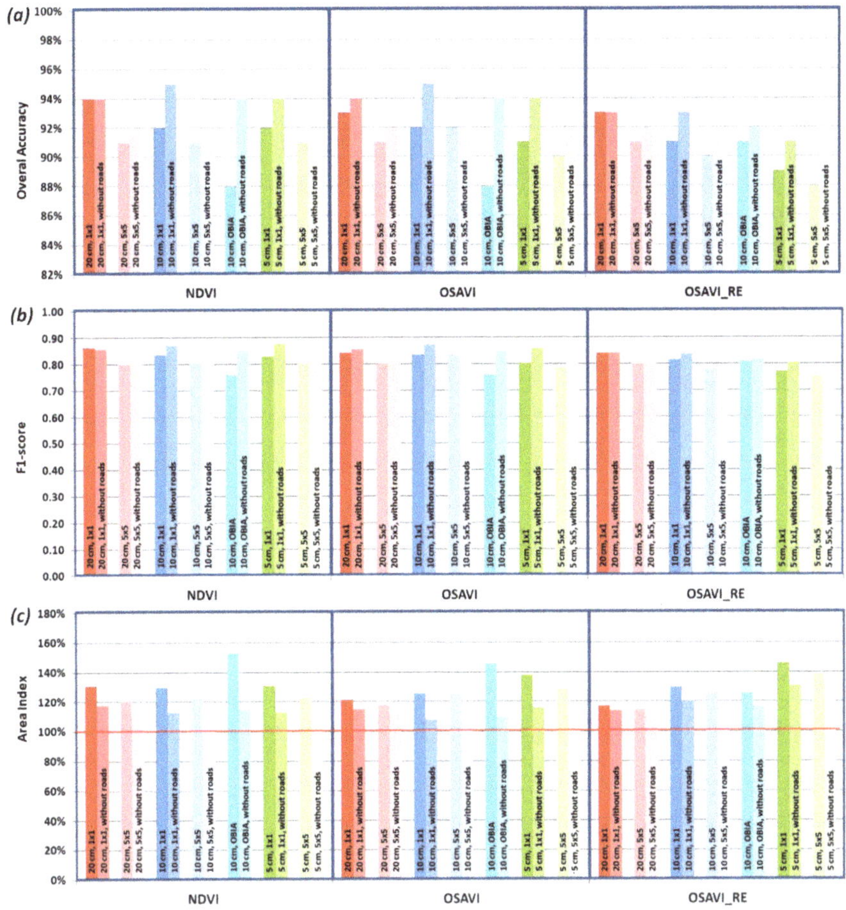

Figure 9. The comparison of the overall accuracy (OA) (**a**), the F1-score (**b**), and area index (AI) (**c**) of all variants of classifications.

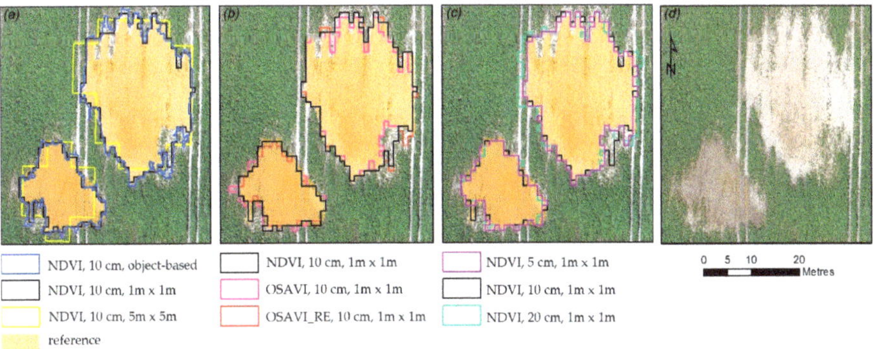

Figure 10. Small fragment of true color orthomosaic with reference mask and detected areas of damaged rapeseed without the influence of roads: (**a**) in three different approaches, (**b**) with three different indices, (**c**) with three different resolutions (**d**) true color orthomosaic.

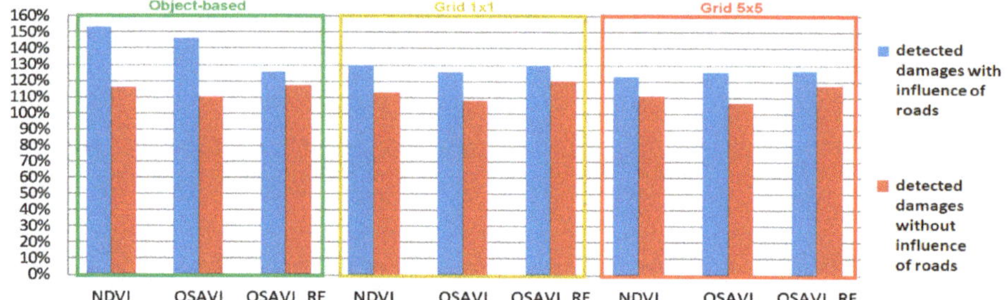

Figure 11. Area index (AI) in three approaches with 10 cm resolution of input data.

In the next section, a deeper analysis of the influence of road removal, the spatial resolution of collected images, and the vegetation indices used in this experiment is provided.

4. Discussion

4.1. Influence of Road Removal and Approaches in Damage Detection

As expected, road removal was a crucial step for obtaining better results. It allowed us to avoid many errors in the classification of smaller objects connected with roads (see Figure 7). It also prevented the incorrect enlargement of detected losses by incorporating road areas in detected polygons.

The negative influence of roads is highlighted in all three approaches on 10 cm input rasters. In both the second (grid 1 × 1 m) and the third (grid 5 × 5 m) approaches, each index increases UA by 3–5 percentage points, with a slight decrease of PA (around three percentage points) with only one exception of a higher outlier (10 percentage points for OSAVI in grid 5 × 5 m) (see Table A1 in Appendix A). The direction of changes in the first approach (object-based) are similar. The decrease of PA is on the same level as in other approaches, however, the increase in UA equals 17 percentage points on OSAVI and NDVI indices, and only two percentage points for OSAVI _RE, which is due to a much bigger UA value for this index before road removal (73–63% and 64% for NDVI and OSAVI, respectively). Although the differences in the range of increase of UA are high, the discrepancies in the final results between approaches 1 and 2 do not exceed three percentage points. This is also reflected by the area index (see Figure 11). The area index's values decrease in all three approaches after using the road removal technique. For OSAVI and NDVI indices, the difference is over two times higher in the object-based approach, but for OSAVI_RE, the decrease is on the same level as in the pixel-based approaches, which may be due to the fact that in the OSAVI_RE variant, technical roads are slightly less visible and are harder to distinguish in the images. This results in a lower number of roads that qualified as winter damage polygons.

The conducted analysis indicates that without removing road influence, object-based classification is a significantly worse method for proper winter damage detection. The generalization that is performed during the process of averaging the cell value in approaches 2 and 3 (resolution of grids 1 and 5 m, respectively), causes the rejection of some fraction of roads, which is not the case in the object-based approach because it provides a more precise classification. Because of this characteristic, a higher resistance to road influence in the OSAVI_RE index is most observable in the object-based scenario.

4.2. Influence of Spatial Resolution

The comparison of PA and UA for a 1 × 1 m grid with respect to vegetation indices is shown in Figure 12. Comparing NDVI results in this approach, we can assume that the smaller the GSD (ground sampling distance of obtained images), the better the results (an increase in the F1-score is observed). However, UA and PA values are almost equal for resolutions of 5 and 10 cm. The visible difference comes only at the 20 cm raster (a decrease of three percentage points in UA and 0.02 in F1-score value). In the case of the OSAVI index, we can confirm that results are slightly better on the 10 cm raster than 5 cm, but 20 cm still results in the worst outcome. The OSAVI_RE index trend is contrary to NDVI, with most of the indicator's values rising with decreasing pixel size. In the second approach, a 20 cm input OSAVI_RE raster results in a F1-score value of 0.84 and an OA equal to 93% (Table A1 in Appendix A), which is only slightly worse than the results obtained from other indices. The detected damage area is also more similar to the reference data than in comparable variants. This information allows us to suppose that, with an increasing size of GSD, OSAVI_RE could be the best index to use in the whole process. Generally, it can be assumed that any tested spatial resolution of UAV-based imagery is enough for winter damage assessment for insurance purposes and there are few differences in terms of accuracy in the tested approaches in the detection of these areas without plants.

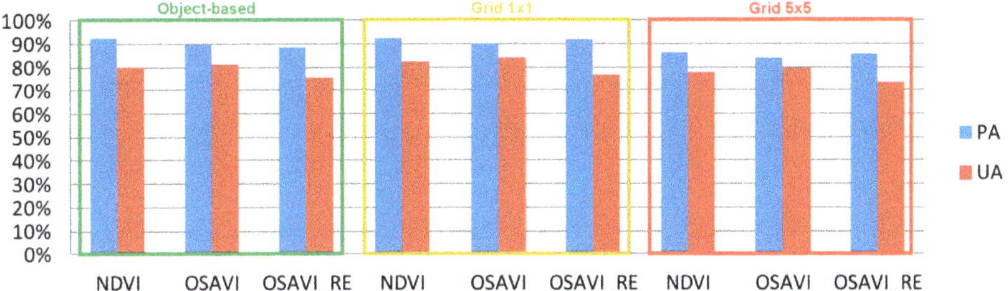

Figure 12. Producer's (PA) and user's (UA) accuracy of the detected damaged crops, without the influence of roads, with three different resolution of input data in the second approach (1 × 1 m).

For the results obtained in the described experiments with NDVI, decreasing the size of GSD provided better results, while the research of Candiago et al. (2015) [9] showed that the use of such a high resolution can be a problem when applying algorithms originally developed for aerial and satellite data. In this paper, three approaches were tested with OBIA, or resampling to grid sizes of 1 × 1 m and 5 × 5 m, which limited this problem.

The highest overall accuracy of detection wais achieved using the NDVI index on two different input data sets with 5 and 20 cm GSD (see Table A1 in Appendix A). For rasters with 10 cm resolution, the better choice was the OSAVI index, which was clearly confirmed in the majority of variants. However, the difference between the overall accuracy of damage detection using NDVI and OSAVI is not significant (1–2%). The usefulness of the NDVI vegetation index to determine the vegetation fraction (the percentage of green vegetation per unit of ground surface) and the number of plants per unit of ground surface is confirmed by the results of other authors (among others [19,22]). In our experiment, the results related to OSAVI_RE index are slightly better.

4.3. *Influence of Vegetation Index Used*

Figure 13 presents the results of damage detection for different vegetation indices in the case of the three approaches. Analysis of this figure allows us to draw a few crucial conclusions. Firstly, we can state that the producer's accuracy is stable in all three approaches regardless of chosen vegetation index as input data for classification. Its values vary from 0 to 3 percentage points in the same approach. However, approach 3 (grid 5 × 5 m) provides smaller absolute PA (around 85%) and UA (77%) values than the other approaches (mean values of 91% and 80%, respectively). This is mostly the result of much stronger generalization in this approach. Secondly, it is observed that there are significant differences in the user's accuracy between the chosen indices in all three approaches.

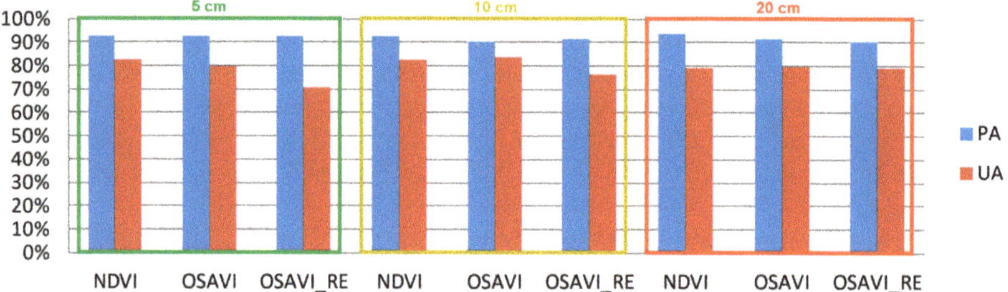

Figure 13. Producer's (PA) and user's (UA) accuracy of results without the influence of roads, with three different approaches performed on 10 cm input data.

In each variant, OSAVI_RE has the lowest value. OSAVI provides slightly better UA than NDVI (difference of 1–2 percentage points), but NDVI compensates this disparity with slightly higher PA values. Analysis of the overall accuracy (see Table A1 in Appendix A) indicates that those two indices are the same, however, the area of damage in all three approaches is 5 percentage points better for OSAVI than NDVI. The difference in the F1-score is only 0.01. Nonetheless, for data with a resolution of 5 cm (for both approaches) and 20 cm (for the third approach), the results indicate that NDVI is slightly better than OSAVI (see Table A1 in Appendix A).

5. Conclusions

The experiments performed demonstrate that low-altitude remote sensing can be used for the assessment of damaged areas in the case of insurance compensation. The case study used multispectral UAV images for the damage detection of rapeseed caused during winter. Data were acquired in April, when plant vegetation starts. This is the most important period in estimating damage since it directly exposes the effect of the winter period on vegetation. The presented methodology includes selecting possible vegetation indices, ground sampling resolutions, and approaches for technical road influence elimination, which can be used as an effective alternative for direct measurements in a field that currently are often used during the refurbishment process.

The analysis of GSD images (in the range of 5–20 cm) demonstrates a relatively small (a few percentage points) influence on the obtained results of winterkill detection. The best approximation of the area of the winter damage to crops was obtained using images with GSD of 10 cm. Results for most variants indicate that there is no need to acquire data with a higher resolution than 10 cm. This is a crucial conclusion for UAV application when the size of the area and the GSD of the images are considered within mission planning

The conclusion of the presented experiment is of great practical importance because cameras with four spectral channels (R, G, B, NIR) are becoming increasingly popular and relatively inexpensive. This provides the possibility to choose various vegetation indices appropriately for the analyzed crop and the purpose of the study related to the monitoring of the crop condition. The best results were obtained using NDVI and OSAVI vegetation indices; however, results from other indices were not significantly worse.

Referring to the utilization of the red edge spectral band, an optimized Soil-Adjusted Vegetation Index with the use of the red edge channel instead of the red channel is the most resilient index to the negative influence of technical roads, which is crucial when road detection cannot be conducted. The optimized soil-adjusted vegetation index with red edge channel (OSAVI_RE) may be the best for crop condition monitoring using lower spatial resolution data.

During tests, it was also necessary to develop and implement elimination of technical roads in calculations of rapeseed area damaged with frost. Including the detected area in the classification helps in the reliable calculation of the area for compensation. Without removing road influence, object-based classification is a significantly worse method, but elimination of road influence can be processed using both object- and pixel-based approaches. In tests with NDVI and OSAVI, overestimation of 10–30% was noticed for calculations without the elimination of the impact of roads, and all of the remote sensing-based approaches in all experiments were characterized by overestimation of at least a few percent considering the Area Index.

The implementation of the presented solutions for delimiting areas for refurbishment in the comprehensive damage estimation method, which also includes periodic monitoring and time analysis, can significantly improve the compensation process, and provide it with scientific certainty for obtained results.

Author Contributions: Conceptualization, K.S., W.O., and K.B.; methodology, W.O.; software, K.S.; validation, W.O.; formal analysis, K.O.-S.; investigation, Ł.J.; resources, K.S. and W.O.; writing—original draft preparation, Ł.J., W.O., K.O.-S., and K.B.; writing—review and editing, K.B. and K.O.-S.; visualization, K.O.-S. and Ł.J.; supervision, project administration and funding acquisition, K.O.-S. and K.B. All authors have read and agreed to the published version of the manuscript.

Funding: The presented research was financed by statutory subsidies of the Polish Ministry of Science and Higher Education for Warsaw University of Technology.

Acknowledgments: We would like to thank Skysnap Sp. z o.o. for providing the data from the Sequoia camera and participation in the calculations and analyzes, and Paulina Bartkowiak for the help in reference data preparation.

Conflicts of Interest: The authors declare no conflict of interest.

Appendix A

Table A1. The results of all variants within the experiment, assessed by producer's accuracy (PA) and user's accuracy (UA) for class damaged crops, and area index (AI), F1-score, and overall accuracy (OA) of classification for damaged crops and healthy crops.

	Resolution			**5 cm**					**5 cm**							
	Approach			**Grid 1 × 1**					**Grid 5 × 5**							
	Index	**UA**	**PA**	**AI**	**F1**	**OA**	**UA**	**PA**	**AI**	**F1**	**OA**					
without influence of roads	**NDVI**	83%	93%	113%	0.88	94%	78%	87%	111%	0.82	92%					
	OSAVI	80%	93%	116%	0.86	94%	76%	87%	114%	0.81	92%					
	OSAVI_RE	71%	93%	131%	0.81	91%	68%	88%	129%	0.77	89%					
with influence of roads	**NDVI**	73%	96%	131%	0.83	92%	73%	90%	123%	0.81	91%					
	OSAVI	69%	96%	138%	0.80	91%	70%	90%	129%	0.79	90%					
	OSAVI_RE	65%	95%	146%	0.77	89%	65%	90%	139%	0.75	88%					
	Resolution			**10 cm**					**10 cm**					**10 cm**		
	Approach			**Grid 1 × 1**					**Grid 5 × 5**					**Object-Based**		
	Index	**UA**	**PA**	**AI**	**F1**	**OA**	**UA**	**PA**	**AI**	**F1**	**OA**	**UA**	**PA**	**AI**	**F1**	**OA**
without influence of roads	**NDVI**	82%	93%	113%	0.87	95%	78%	86%	111%	0.82	92%	80%	92%	115%	0.86	94%
	OSAVI	84%	91%	108%	0.87	95%	79%	84%	106%	0.81	92%	81%	90%	110%	0.85	94%
	OSAVI_RE	77%	92%	120%	0.84	93%	73%	86%	117%	0.79	91%	76%	89%	117%	0.82	92%
with influence of roads	**NDVI**	74%	96%	130%	0.84	92%	73%	90%	123%	0.81	91%	63%	96%	153%	0.76	88%
	OSAVI	75%	94%	126%	0.83	92%	75%	94%	126%	0.83	92%	64%	94%	146%	0.76	88%
	OSAVI_RE	72%	94%	130%	0.82	91%	70%	88%	126%	0.78	90%	73%	91%	126%	0.81	91%
	Resolution			**20 cm**					**20 cm**							
	Approach			**Grid 1 × 1**					**Grid 5 × 5**							
	Index	**UA**	**PA**	**AI**	**F1**	**OA**	**UA**	**PA**	**AI**	**F1**	**OA**					
without influence of roads	**NDVI**	79%	94%	118%	0.86	94%	78%	86%	111%	0.82	92%					
	OSAVI	80%	92%	115%	0.86	94%	76%	86%	113%	0.81	92%					
	OSAVI_RE	79%	90%	114%	0.84	93%	76%	85%	112%	0.80	92%					
with influence of roads	**NDVI**	76%	100%	131%	0.86	94%	73%	89%	121%	0.80	91%					
	OSAVI	77%	93%	121%	0.84	93%	74%	87%	118%	0.80	91%					
	OSAVI_RE	78%	91%	117%	0.84	93%	75%	86%	115%	0.80	91%					

References

1. EUROSTAT DATA. 2019. Available online: https://ec.europa.eu/eurostat/databrowser/view/tag00100/default/table?lang=en (accessed on 17 July 2020).
2. GUS. Statistical Yearbook of Agriculture. Warsaw. 2020. Available online: https://stat.gov.pl (accessed on 17 July 2020).
3. Szulc, K. Wymarzanie Rzepaku Ozimego, Farmer.pl. 2018. Available online: https://www.farmer.pl/produkcja-roslinna/rosliny-oleiste/wymarzanie-rzepaku-ozimego,76896.html (accessed on 17 July 2020).
4. Wójtowicz, M.; Wójtowicz, A.; Piekarczyk, J. Application of remote sensing methods in agriculture. *Commun. Biometry Crop Sci.* **2016**, *11*, 31–50.
5. Brown, R.J.; Staenz, K.; McNairn, H.; Hopp, B.; Van Acker, R. Application of high-resolution optical imagery to precision agriculture. In Proceedings of the International Symposium Geomatics in the Era of RADARSAT (GER'97), Ottawa, ON, Canada, 25–30 May 1997; p. 9.
6. Haboudane, D.; Miller, J.R.; Tremblay, N.; Zarco-Tejada, P.J.; Dextraze, L. Integrated narrow-band vegetation indices for prediction of crop chlorophyll content for application to precision agriculture. *Remote Sens. Environ.* **2002**, *81*, 416–426. [CrossRef]
7. Doraiswamy, P.C.; Moulin, S.; Cook, P.W.; Stern, A. Crop yield assessment from remote sensing. *Photogramm. Eng. Remote Sens.* **2003**, *69*, 665–674. [CrossRef]
8. Rembold, F.; Atzberger, C.; Savin, I.; Rojas, O. Using low resolution satellite imagery for yield prediction and yield anomaly detection. *Remote Sens.* **2013**, *5*, 1704–1733. [CrossRef]
9. Candiago, S.; Remondino, F.; De Giglio, M.; Dubbini, M.; Gattelli, M. Evaluating multispectral images and vegetation indices for precision farming applications from UAV images. *Remote Sens.* **2015**, *7*, 4026–4047. [CrossRef]
10. Gago, J.; Douthe, C.; Coopman, R.; Gallego, P.; Ribas-Carbo, M.; Flexas, J.; Medrano, H. UAVs challenge to assess water stress for sustainable agriculture. *Agric. Water Manag.* **2015**, *153*, 9–19. [CrossRef]
11. Holman, F.H.; Riche, A.B.; Michalski, A.; Castle, M.; Wooster, M.J.; Hawkesford, M.J. High throughput field phenotyping of wheat plant height and growth rate in field plot trials using UAV based remote sensing. *Remote Sens.* **2016**, *8*, 1031. [CrossRef]
12. Puliti, S.; Ørka, H.O.; Gobakken, T.; Næsset, E. Inventory of Small Forest Areas Using an Unmanned Aerial System. *Remote Sens.* **2015**, *7*, 9632–9654. [CrossRef]
13. Lucieer, A.; Jong, S.M.D.; Turner, D. Mapping landslide displacements using Structure from Motion (SfM) and image correlation of multi-temporal UAV photography. *Prog. Phys. Geogr.* **2014**, *38*, 97–116. [CrossRef]
14. Wei, Z.; Han, Y.; Li, M.; Yang, K.; Yang, Y.; Luo, Y.; Ong, S.-H. A Small UAV Based Multi-Temporal Image Registration for Dynamic Agricultural Terrace Monitoring. *Remote Sens.* **2017**, *9*, 904. [CrossRef]
15. Khot, L.R.; Sankaran, S.; Carter, A.H.; Johnson, D.A.; Cummings, T.F. UAS imaging-based decision tools for arid winter wheat and irrigated potato production management. *Int. J. Remote Sens.* **2016**, *37*, 125–137. [CrossRef]
16. Kipp, S.; Mistele, B.; Baresel, P.; Schmidhalter, U. High-throughput phenotyping early plant vigour of winter wheat. *Eur. J. Agron.* **2014**, *52*, 271–278. [CrossRef]
17. Zhou, J.; Pavek, M.J.; Shelton, S.C.; Holden, Z.J.; Sankaran, S. Aerial multispectral imaging for crop hail damage assessment in potato. *Comput. Electron. Agric.* **2016**, *127*, 406–412. [CrossRef]
18. Salamí, E.; Barrado, C.; Pastor, E. UAV flight experiments applied to the remote sensing of vegetated areas. *Remote Sens.* **2014**, *6*, 11051–11081. [CrossRef]
19. Shi, Y.; Thomasson, J.A.; Murray, S.C.; Pugh, N.A.; Rooney, W.L.; Shafian, S.; Rajan, N.; Rouze, G.; Morgan, C.L.S.; Neely, H.L.; et al. Unmanned Aerial Vehicles for High-Throughput Phenotyping and Agronomic Research. *PLoS ONE* **2016**, *11*, e0159781. [CrossRef] [PubMed]
20. Zaman-Allah, M.; Vergara, O.; Araus, J.L.; Tarekegne, A.; Magorokosho, C.; Zarco-Tejada, P.J.; Hornero, A.; Hernández Albà, A.; Das, B.; Craufurd, P.; et al. Unmanned aerial platform-based multi-spectral imaging for field phenotyping of maize. *Plant Methods* **2015**, *11*, 1–10. [CrossRef] [PubMed]
21. Yang, G.; Liu, J.; Zhao, C.; Li, Z.; Huang, Y.; Yu, H.; Xu, B.; Yang, X.; Zhu, D.; Zhang, X.; et al. Unmanned Aerial Vehicle Remote Sensing for Field-Based Crop Phenotyping: Current Status and Perspectives. *Front. Plant Sci.* **2017**, *8*, 1–26. [CrossRef]

22. Ni, J.; Yao, L.; Zhang, J.; Cao, W.; Zhu, Y.; Tai, X. Development of an Unmanned Aerial Vehicle-Borne Crop-Growth Monitoring System. *Sensors* **2017**, *17*, 502. [CrossRef]
23. Shanahan, J.F.; Schepers, J.S.; Francis, D.D.; Varvel, G.E.; Wilhelm, W.W.; Tringe, J.M.; Schlemmer, M.R.; Major, D.J. Use of Remote-Sensing Imagery to Estimate Corn Grain Yield. *Agron. J.* **2001**, *93*, 583–589. [CrossRef]
24. Zhang, C.; Walters, D.; Kovacs, J.M. Applications of Low Altitude Remote Sensing in Agriculture upon Farmers' Requests—A Case Study in Northeastern Ontario, Canada. *PLoS ONE* **2014**, *9*, e112894. [CrossRef]
25. Hunt, E.R.; Hively, W.D.; Daughtry, C.S.; McCarty, G.W.; Fujikawa, S.J.; Ng, T.L.; Yoel, D.W. Remote sensing of crop leaf area index using unmanned airborne vehicles. In Proceedings of the Pecora 17 Symposium, Denver, CO, USA, 18–20 November 2008.
26. Bochenek, Z.; Dąbrowska-Zielińska, K.; Ciołkosz, A.; Drupka, S.; Boken, V.K. *Monitoring Agricultural Drought in Poland. Monitoring and Predicting Agricultural Drought*; Oxford University Press: Oxford, UK, 2005; pp. 171–180.
27. Piekarczyk, J.; Sulewska, H.; Szymańska, G. Winter oilseed-rape yield estimates from hyperspectral radiometer measurements. *Quaest. Geogr.* **2011**, *30*, 77–84. [CrossRef]
28. Hunt, E.R.; Hively, W.D.; Fujikawa, S.J.; Linden, D.S.; Daughtry, C.S.T.; McCarty, G.W. Acquisition of NIR-Green-Blue Digital Photographs from Unmanned Aircraft for Crop Monitoring. *Remote Sens.* **2010**, *2*, 290–305. [CrossRef]
29. Mathews, A.J.; Jensen, J.L.R. Visualizing and quantifying vineyard canopy LAI using an unmanned aerial vehicle (UAV) collected high density structure from motion point cloud. *Remote Sens.* **2013**, *5*, 2164–2183. [CrossRef]
30. Wei, C.; Huang, J.; Wang, X.; Blackburn, G.A.; Zhang, Y.; Wang, S.; Mansaray, L.R. Hyperspectral characterization of freezing injury and its biochemical impacts in oilseed rape leaves. *Remote Sens. Environ.* **2017**, *195*, 56–66. [CrossRef]
31. Jones, H.G.; Vaughan, R.A. *Remote Sensing of Vegetation—Principles, Techniques, and Applications*; Oxford University Press: New York, NY, USA, 2010.
32. Nicotra, A.B.; Hofmann, M.; Siebke, K.; Ball, M.C. Spatial patterning of pigmentation in evergreen leaves in response to freezing stress. *Plant Cell Environ.* **2003**, *26*, 1893–1904. [CrossRef]
33. Wu, Q.; Zhu, D.; Wang, C.; Ma, Z.; Wang, J. Diagnosis of freezing stress in wheat seedlings using hyperspectral imaging. *Biosyst. Eng.* **2012**, *112*, 253–260. [CrossRef]
34. Flower, K.; Boru, B.; Nansen, C.; Jones, H.; Thompson, S.; Lacoste, C.; Murphy, M. *Proof of Concept: Remote Sensing Frost-Induced Stress in Wheat Paddocks*; Grains Research and Development Corporation: Canberra, Australia, 2014.
35. Murphy, M.E.; Boruff, B.; Callow, J.N.; Flower, K.C. Detecting Frost Stress in Wheat: A Controlled Environment Hyperspectral Study on Wheat Plant Components and Implications for Multispectral Field Sensing. *Remote Sens.* **2020**, *12*, 477. [CrossRef]
36. Xie, Y.; Wang, C.; Yang, W.; Feng, M.; Qiao, X.; Song, J. Canopy hyperspectral characteristics and yield estimation of winter wheat (Triticum aestivum) under low temperature injury. *Sci. Rep.* **2020**, *10*, 244. [CrossRef]
37. Gitelson, A.A.; Kaufman, Y.J.; Merzlyak, M.N. Use of a green channel in remote sensing of global vegetation from EOS-MODIS. *Remote Sens. Environ.* **1996**, *58*, 289–298. [CrossRef]
38. Daughtry, C.S.T.; Walthall, C.L.; Kim, M.S.; Brown de Colstoun, E.; McMurtrey, J.E., III. Estimating corn leaf chlorophyll concentration from leaf and canopy reflectance. *Remote Sens. Environ.* **2000**, *74*, 229–239. [CrossRef]
39. Rouse, J.W. *Monitoring the Vernal Advancement and Retrogradation (Green Wave Effect) of Natural Vegetation*; Technical Report; NASA: Washington, DC, USA, 1972.
40. Huete, A.R. A soil-adjusted vegetation index (SAVI). *Remote Sens. Environ.* **1988**, *25*, 295–309. [CrossRef]
41. Rondeaux, C.; Steven, M.D.; Baret, F. Optimized of Soil-adjusted vegetation indices. *Remote Sens Environ.* **1996**, *55*, 95–107. [CrossRef]
42. Schuster, C.; Förster, M.; Kleinschmit, B. Testing the red edge channel for improving land-use classifications based on high-resolution multi-spectral satellite data. *Int. J. Remote Sens.* **2012**, *33*, 5583–5599. [CrossRef]

43. Li, F.; Miao, Y.; Feng, G.; Yuan, F.; Yue, S.; Gao, X.; Liu, Y.; Liu, B.; Ustin, S.L.; Chen, X. Improving estimation of summer maize nitrogen status with red edge-based spectral vegetation indices. *Field Crop. Res.* **2014**, *157*, 111–123. [CrossRef]
44. Congalton, R. A Review of Assessing the Accuracy of Classifications of Remotely Sensed Data. *Remote Sens. Environ.* **1991**, *37*, 35–46. [CrossRef]
45. Koga, Y.; Miyazaki, H.; Shibasaki, R. A CNN-based Method of vehicle detection from aerial images using hard example mining. *Remote Sens.* **2018**, *10*, 124. [CrossRef]

Article

Automatic Grapevine Trunk Detection on UAV-Based Point Cloud

Juan M. Jurado [1,*], Luís Pádua [2,3], Francisco R. Feito [1] and Joaquim J. Sousa [2,3]

1. Computer Graphics and Geomatics Group of Jaén, University of Jaén, 23071 Jaén, Spain; ffeito@ujaen.es
2. Engineering Department, School of Science and Technology, University of Trás-os-Montes e Alto Douro, 5000-801 Vila Real, Portugal; luispadua@utad.pt (L.P.); jjsousa@utad.pt (J.J.S.)
3. Centre for Robotics in Industry and Intelligent Systems (CRIIS), INESC Technology and Science (INESC-TEC), 4200-465 Porto, Portugal

* Correspondence: jjurado@ujaen.es

Received: 26 August 2020; Accepted: 16 September 2020; Published: 17 September 2020

Abstract: The optimisation of vineyards management requires efficient and automated methods able to identify individual plants. In the last few years, Unmanned Aerial Vehicles (UAVs) have become one of the main sources of remote sensing information for Precision Viticulture (PV) applications. In fact, high resolution UAV-based imagery offers a unique capability for modelling plant's structure making possible the recognition of significant geometrical features in photogrammetric point clouds. Despite the proliferation of innovative technologies in viticulture, the identification of individual grapevines relies on image-based segmentation techniques. In that way, grapevine and non-grapevine features are separated and individual plants are estimated usually considering a fixed distance between them. In this study, an automatic method for grapevine trunk detection, using 3D point cloud data, is presented. The proposed method focuses on the recognition of key geometrical parameters to ensure the existence of every plant in the 3D model. The method was tested in different commercial vineyards and to push it to its limit a vineyard characterised by several missing plants along the vine rows, irregular distances between plants and occluded trunks by dense vegetation in some areas, was also used. The proposed method represents a disruption in relation to the state of the art, and is able to identify individual trunks, posts and missing plants based on the interpretation and analysis of a 3D point cloud. Moreover, a validation process was carried out allowing concluding that the method has a high performance, especially when it is applied to 3D point clouds generated in phases in which the leaves are not yet very dense (January to May). However, if correct flight parametrizations are set, the method remains effective throughout the entire vegetative cycle.

Keywords: grapevine detection; precision viticulture; 3D vineyard structure; photogrammetry

---

1. Introduction

The monitoring and management of agricultural crops, particularly with regard to nutrient level, water stress, diseases and pests, and phenological status, are vital for successful agricultural operations [1]. Traditionally, these activities are carried out through visual examinations of the crops, or by analysing plants and soil, which are time-consuming and invasive approaches [2]. Considering the fact that it is necessary to maximise yield and resources, while reducing environmental impacts, mainly by optimising the use of water and significantly reducing fertilisers and pesticides [3]. This can only be achieved by obtaining data that allow the intelligent and sustainable management of agricultural parcels [4]. It will then be possible, in a rational and economical way, to resort differentiated and localised actions with regard to the use of water and nutrients, and to control the soil and vegetation cover, as well as the plant's phytosanitary status.

The technological advances in recent years, have enabled the miniaturisation of electronic components and a significantly reduction in prices, taking Precision Agriculture (PA) to another level. For example, the advent of Unmanned Aerial Vehicles (UAVs) capable of capturing aerial high resolution data using different kind of sensors (RGB, multi and hyperspectral, thermal and LiDAR), together with new photogrammetric processing methods, allow the computation of diverse outcomes such as orthophoto mosaics, vegetation indices and 3D point clouds [5]. UAVs are a popular tool in PA and the obtained aerial imagery is turned into information which can be used to optimise crop inputs through variable rate applications [6–8].

In a short period of time PA approaches and practices have become very popular and were introduced in all agricultural sectors [9]. The vine and wine sector is among those that have most benefited of precision farming techniques, applied to optimise vineyard performance [10]. Thus, the Precision Viticulture (PV) concept was introduced and can be defined as a particular field of PA, whose purpose is maximising grape yield and quality while minimising environmental impacts and risks [11]. Therefore, it is possible to avoid unnecessary treatments, which can be harmful and polluting, and reduce costs [12]. The ability of UAVs to obtain high spatiotemporal resolution and geocoded images from different sensors, make them a powerful tool for a better understanding of the spatial and multi-temporal heterogeneity of vineyard plots, allowing the estimation of parameters that directly affect its state. Thus, individual grapevine identification and location is of great importance to precisely assess the vineyard status estimating different parameters per individual plant [13]. However, there are many features in vineyards that make these scenarios very complex to develop automatic methods for trunk individual detection and location [14]. Therefore, segmentation methods that consist of processes of dividing input data into several disjoint areas that maintain the unique and homogeneous features from surrounding have to be employed.

Regarding vineyard vegetation detection, several methods were already proposed based on different approaches using the photogrammetric outcomes from UAV-based imagery by applying image processing techniques, machine learning methods and by filtering dense 3D point clouds and Digital Elevation Models (DEMs) [14–18]. Those methods are capable of distinguishing grapevine from non-grapevine vegetation and to extract different vineyard macro properties such as the number of vine rows, row spacing, width and height, potential missing plants and vineyard vigour maps.

The outcomes resulting from photogrammetric processing applied to UAV-based imagery can be used to estimate individual geometrical and biophysical grapevine parameters, providing a plant-specific application for PV [19]. In this scope some studies can be found in the literature. De Castro et al. [20] developed an Object-Based Image Analysis (OBIA) method applied to high-resolution vineyard Digital Surface Models (DSMs) to estimate grapevine vegetation. Then, the individual position of grapevines were marked, assuming a constant space between plants. This way, missing plants were also estimated and some geometrical parameters were estimated. In a different study, proposed by Matese and Di Genaro [21], missing plants detection was assessed in a semi-automatic procedure by filtering the DSM and by manually placing small polygons, representing individual plants and, then, analysing the number of pixels intercepted by each polygon by using a five-classes approach based on quantiles to verify the probability of a missing plant presence. A binary multivariate-logistic regression model was used by Primicerio et al. [22] for the individual detection of grapevines, including missing grapevines, in orthophoto mosaics. In the referred studies it is highlighted that the integration of other sensors data could allow the extraction of single plant vigour, health and water status. In this regard, Pádua et al. [13] performed an individual grapevine estimation for site-specific management in a multi-temporal context, helping winegrowers to fully explore the potential of the high-resolution data provided by UAVs and to combine data resultant from the different imagery sensors for a more precise decision support and a quick vineyard inspection. More recently, several studies have explored 3D point clouds resulting from UAV-based imagery photogrammetry processing to identify vineyards. Point cloud models consist of large datasets of points representing the surface of visible objects and can be derived from UAV-based imagery by photogrammetry and computer vision algorithms such as, for example, Structure from Motion (SfM). Alternatively, 3D point clouds can be directly provided by Light Detection and Ranging Systems (LiDAR). Comba et al. [23] proposed an unsupervised algorithm for vineyard detection and vine-rows features evaluation, based on 3D point-cloud maps processing. However, as final result, only the vineyards and local evaluation of vine rows orientation were retrieved. Comba et al. [24] applied a multivariate linear regression model to crop canopy descriptors derived from the 3D point cloud, to estimate vineyard's Leaf Area Index (LAI). Marie Weiss and Frédéric Baret [17], applied a SfM algorithm to extract 3D dense point cloud over the vineyard and used the terrain altitude, extracted from the dense point cloud, to get the 2D distribution of height of the vineyard. Then, a threshold on the height was applied to separate the rows. Mesas-Carrascosa et al. [25] used 3D point clouds generated using photogrammetric techniques to RGB images acquired by UAV to derive vineyard canopy information. Additionally, to the geometry, each 3D point also stored the colour which was used to discriminate between vegetation and bare soil. Aboutalebi et al. [26] used UAV-based 3D information to monitor and assess vineyard plant's condition. Different aspects of 3D point cloud were used to estimate height, volume, surface area, and projected surface area of plant's canopy. Then biomass information was used to assess its relationship with in situ LAI. Other studies, such as that by Moreno et al. [27], used terrestrial LiDAR sensors to reconstructed vineyard crops. Although accurate, these methods are time-consuming and very expensive.

As it can be concluded, through the studies previously presented, there are many groups of researchers who are dedicated to the development of methods to extract useful information from vineyards. Although it is considered by everyone of fundamental importance the detection and location of individual plants, there are no methods capable of making it fully automatic. Indeed, the various methods found in the literature, are able to estimate the position of trunks, but using prior knowledge related to the number of plants per row and the distance between plants. Therefore, a fully automatic method able to detect and locate grapevine trunks is desirable and would have the potential to create base maps for most PV studies.

In this article, we present an innovative and fully automatic method able to detect and locate individual grapevine trunks, by exploring 3D point clouds derived from photogrammetric processing of UAV-based RGB imagery. The proposed method proved to be effective even when applied to complex vineyards plots. It is able to distinguish posts from trunks and to mark missing plants.

2. Materials and Methods

2.1. Study Area

To develop the method proposed in this manuscript, several commercial vineyards (Figure 1a), in the northern region of Portugal, were selected to its application. Commercial vineyards usually present the great advantage of a proper management, where the best practices are applied to improve yield and quality. Therefore, these vineyards present well treated rows with a regular vegetative wall, facilitating the individual grapevine detection. Then, in the scope of this study, a complex and challenging vineyard plot was analysed (Figure 1b, 41°17′08.1″ N, 7°44′09.9″ W, 472 m altitude). The main purpose of using this vineyard, located in the campus of the University of Trás-os-Montes e Alto Douro (UTAD), Vila Real, Portugal, in the Douro demarcated region, was to push the method's application to its limit. The plot has an area of 3200 m^2 and is composed of 55 rows, and a double guyot trained system is used. The selection of this vineyard is based on the different levels of vigour and missing plants along the vine rows, providing a diverse variety of cases that are hard to be found in commercial vineyards.

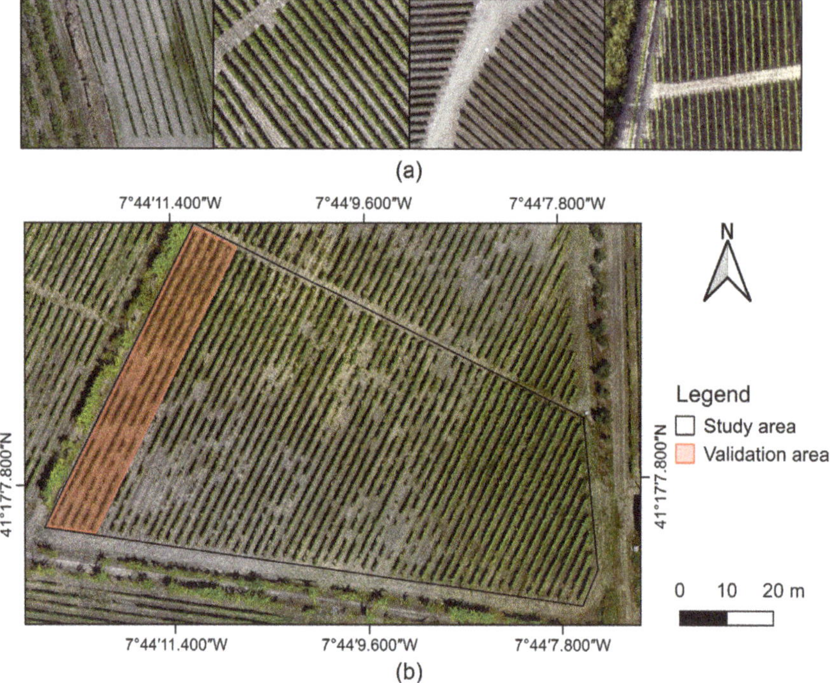

Figure 1. General overview of the study areas: (**a**) some examples of commercial vineyards; and (**b**) vineyard plot used for validation and to assess the limits of the proposed methodology. Coordinates in WGS84 (EPSG:4326).

2.2. UAV-Based Data Acquisition

Aerial RGB imagery acquisition was performed using the multi-rotor UAV DJI Phantom 4 (DJI, Shenzhen, China). Its native camera was used for RGB imagery acquisition, FCC 3 model, a CMOS sensor with 12.4 MP resolution mounted in a 3-axis electronic gimbal.

Different flights were conducted over distinct areas using a single-grid configuration and a flight height varying between 30 m (June and July flight campaigns) and 50 m (flights from January to May), from the UAV take-off position and with an imagery frontal overlap of 90% and 80% side overlap. The missions were planned and executed using DroneDeploy (California, CA, USA) in an Android smartphone. Regarding the most complex test site (Figure 1b), the flight was performed on 30 July 2019 and a total of 228 images were acquired. The whole flight campaign was carried in 13 min, five minutes for UAV assembly/disassembly operations and mission uploading, while the duration of the flight was eight minutes. The camera can be facing down, i.e., in the nadir direction, in all the flights conducted during the season preparation period. In the remaining flights, conducted in the growing and/or harvesting preparation period, the camera was used with an inclination angle of 65°, relative to the nadir direction. This choice was done due to the absence or presence of leaves, capable of obstructing trunk detection (respectively, Phase 1 and Phases 2 and 3, of Figure 2).

Figure 2. The three moments of the vegetative cycle that influence camera parameterization and flight height. Phase 1 (beginning of the wine campaign), where the influence of leaves is negligible—camera can be used facing down and the flight altitude can be higher; Phase 2 (critical phase of phenological development); and Phase 3 (preparation of the vintage and estimating production)—In these phases the camera should be used with an angle and the flight height must be low (20–30 m).

2.3. Proposed Method

The main steps of the automatic grapevine trunk detection method based on a geometric segmentation on point cloud data are presented in Figure 3.

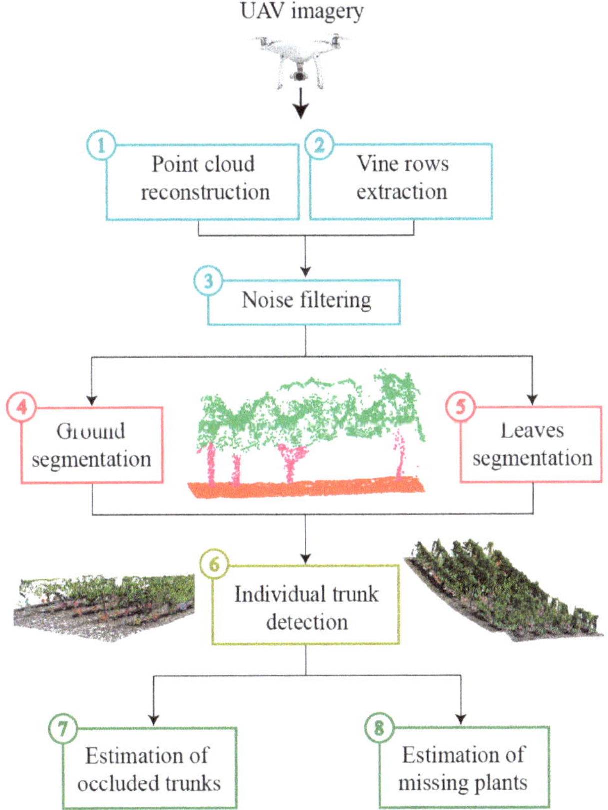

Figure 3. The flowchart diagram of the proposed methodology.

Firstly, the RGB aerial imagery are used to generate a 3D dense point cloud which was geometrically corrected using ground control points (GCPs). Then, a noise reduction is applied to remove many points which usually are close to the plant body. This step is important to achieve a more accurate trunk detection on areas with dense vegetation. Secondly, the location of vine rows is obtained in the form of lines using the method proposed in Pádua et al. [14]. In this way, the search area is confined to the vine rows only, optimising the whole process. Consequently, 3D points close to these lines were selected from the rest of the point cloud. Thirdly, a geometric segmentation is carried out in order to remove 3D points representing the ground and leaves. Thus, points belonging to trunks are isolated and spatial clustering can be performed. Finally, the 3D position for each grapevine is determined and, therefore, the number of existing and missing plants is calculated. The method was developed and implemented in C++ using the Point Cloud Library (PCL) [28].

2.3.1. SFM Reconstruction and Noise Removal

SfM techniques [29] are widely used for 3D reconstruction of multiple scenarios of the real-world [30]. These image-based methods are able to identify and match key points between overlapping images. In contrast to LiDAR-based solutions [27], the application of SfM enables collecting the fit-for-purpose data to model the geometry from some viewpoints. In general, plant modelling poses some challenges due to irregular surfaces, occlusion and varying illumination. In this regard, some considerations should be taken into account to process data correctly. For this study, 3D dense point clouds were generated over vineyard plots, by considering the following processing options: (1) a high overlapping rate ($\geq$80%), (2) a valid key point must be visible in at least three images and (3) the image scale is set to 1/2 in order to increase recognisable key points per image. The photogrammetric processing of the acquired RGB imagery was performed using Pix4Dmapper Pro (Pix4D SA, Lausanne, Switzerland). A 3D dense point cloud was generated using the multi-scale half-image size, a high point density and a minimum of three matches per image. It was exported in polygon file format (PLY). Moreover, raster products were also computed after 3D point cloud interpolation using Inverse Distance Weighting (IDW).

Noisy points that inevitably surround the vegetation areas have to be removed in order to make an accurate geometric segmentation on the 3D model. The point cloud is filtered applying a noise filter which is provided by PCL, this method is based on the computation of distance between neighbours [31]. For each 3D point, the mean distance from it to all its neighbours is computed. Thus, all points whose mean distances are outside an interval defined by the global distances mean and standard deviation are considered to be outliers. The neighbour search was developed by considering a specific radius which is related to the point cloud density. The several tests performed allowed to conclude that 0.05 m should be used to increase method's and results' quality. By applying this noise filter to the 3D point cloud, most of erroneous 3D points in the lower parts of the grapevines were removed, allowing a better recognition of the trunks.

2.3.2. Vine Rows Extraction

In order to reduce the research area, the method proposed by Pádua et al. [14] is first applied. In this way, the vector lines representing individual axis of vine rows are identified. In short, the identification of the lines is based on the use of a crop surface model—computed from the subtraction of the DEM to the Digital Surface Model (DSM)—in combination with the green percentage index (G%) [32], computed using the red, green and blue bands of the orthophoto mosaic. Then, grapevine vegetation is estimated from the threshold and concatenation of both raster products— the Canopy Surface Model (CSM), according to a height range and the G%, by using the Otsu's method [33]. After this procedure, a binary image is generated with a set of clusters of pixels mostly representing the grapevine vegetation. In this way, the vine rows and its central lines are estimated, considering the orientation of the most representative clusters.

Thus, the estimated vine rows axis (central lines) are used as a virtual guide to create a buffer allowing identifying points which compose the plant geometry and the surrounding soil (see Figure 4a). Since the vegetative wall width of grapevine plants usually varies between 30 and 50 cm, it was decided to create a 60 cm buffer to selected 3D points to be analysed. As it is shown in Figure 4b, the green points represent the selected points considering 30 cm width to each side the vine row.

(a) (b)

Figure 4. Example of a segmentation of vine rows: (**a**) visualisation of the 3D vine rows generated as presented in Pádua et al. [14]; and (**b**) 3D points selection in the point cloud, using a 60 cm buffer.

2.3.3. Ground and Leaves Segmentation

The vine rows extraction method enables removing 3D points, which were outside of 3D buffers. However, in addition to points, which potentially may define the trunk's geometry, there are other 3D points in the vine row space to be discarded. In order to isolate trunk points, it is necessary to remove those 3D points representing leaves, which usually appear in the upper section of the point cloud and ground points, which are located in the lower part of the 3D model. For this purpose, only geometric and spatial features as well as the point colour were considered, and the following three steps strategy was applied: (1) spatial subdivision of the vine row buffers, based on height thresholds, (2) ground removal (3) leaves removal. This process is fully automatic and no human intervention is required. In effect, the method has the ability to be applied even in vineyard plots with irregular slope, distinct density of plant foliage and voids (missing plants) along the vine row.

However, and before the application of the this procedure, 3D buffers need to be divided in different segments (Figure 5). This subdivision is determined based on the buffer's length and the terrain's slope. In this task, it is crucial to apply height thresholds, mainly important if the terrain slope varies. In flat terrains this step could be avoided, still to keep the method as general as possible, it was decided to include this step. If the terrain's slope is irregular, a higher number of segments will be required to allow a better fit. By default, the segment length is set to 1 m since it proved to fit most scenarios.

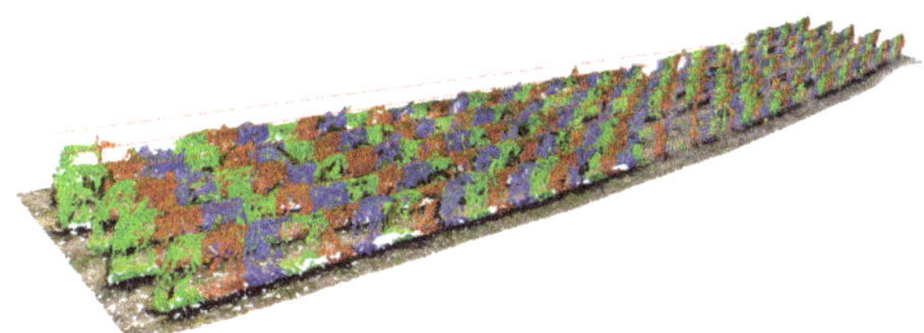

Figure 5. Subdivision of the vine row into n segments.

The leaves and ground removal operation is carried out by considering the vine row buffers segmentation. In fact, the following geometric operations were developed for each segment. Firstly, 3D points with the highest and lowest heights are detected. Then, the terrain's slope is fitted by changing the orientation of a cutting plane for n iterations. Thus, ground points which are under this plane were automatically discarded. Figure 6 shows the main iterations of this step.

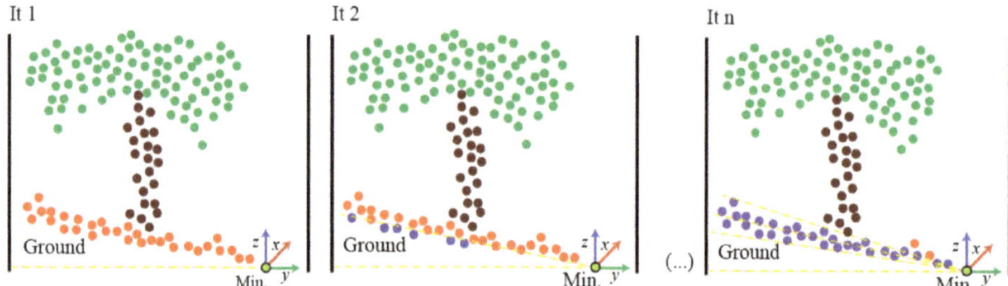

Figure 6. Cutting plane adjustment for the ground removal in segment n_1.

Initially, this 3D plane is fixed by the point with the minimum height and the up vector (**v** = 0,0,1), whose direction is perpendicular to the horizontal plane. Then, for each iteration, the plane is rotated 5° around the x-axis. This geometric transformation was performed by applying the rotation matrix showed in Equation (1). The stopping criterion was determined by several 3D points, selected in each iteration, i.e., those which were under the cutting plane influence. The plan inclination angle is set to the value allowing a proper fit of 3D plane to the terrain slope. Finally, to determine the location of points relative to the cutting plane, the point-normal equation for the cutting plane was applied. If the result is lower than zero, the point is under the plane (Equation (2)), so it is classified as ground and automatically removed in the point cloud.

$$R_x = \begin{pmatrix} 1 & 0 & 0 \\ 0 & \cos(\alpha) & -\sin(\alpha) \\ 0 & \sin(\alpha) & \cos(\alpha) \end{pmatrix} \quad (1)$$

$$A(x - x1) + B(y - y_1) + C(z - z_1) < 0 \quad (2)$$

where A, B and C represent the coefficients of the normal vector, (x, y, z) represent the coordinates of the point on the plane and (x_1, y_1, z_1) represent the coordinates of 3D points.

Regarding the leaves removal, the same inclination angle used for the ground cutting plane was applied. To fix the position of this upper plane, an offset value was calculated considering points with the maximum and minimum height for each segment. This procedure is illustrated in Figure 7.

After ground and leaves points removal, the next step consists in the identification of individual trunks. This is done based on the application of a spatial segmentation especially developed for that purpose. In this regard, the trunk is considered to be a 3D geometric shape, and a clustering method sharing the following features was developed: (1) the Euclidean distance between the 3D points must be lower than 50 cm (this value is estimated considering the typical distance between grapevines, which is never of this order); and (2) the minimum number of points for clustering is set to five. According to these constraints, a correct limitation of the growing region was determined for each cluster. Therefore, n groups of points were segmented for each vine row.

The value h represents the vertical height of plants in the segment, while the value f represents the offset obtained by applying Equation (3). According to this setting, the top plane was adapted by the geometric features for each segment. In this case, upper points to the cutting planes, which were considered vegetation, were removed in the point cloud. However, some outlying points could not be correctly filtered by the cutting plane. For this reason, another filter was applied based on the point colour. A threshold for the green channel was fixed in order to remove vegetation points characterised by the green channel higher than 120 as well as red and blue channel lower than 80. This combination was considered adequate and derived from the many tests performed. Lower values would cause the removal of points belonging to the trunks.

$$f = |Z_{Max} - Z_{Min}|x0.6 \tag{3}$$

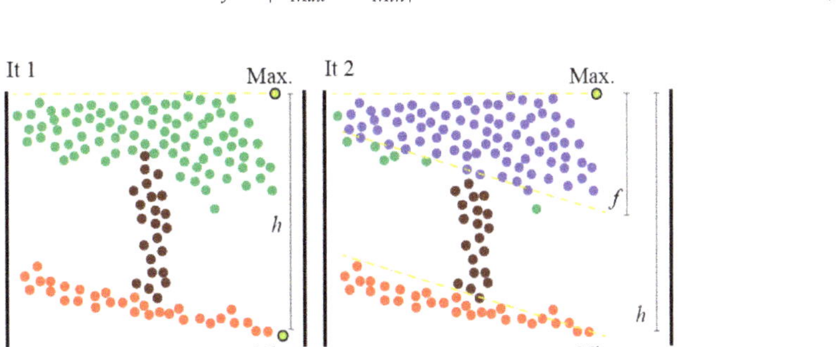

Figure 7. Cutting plane adjustment for the leaves removal.

2.3.4. Trunk Detection

After ground and leaves points removal, the next step consists in the identification of individual trunks. This is done based on the application of a spatial segmentation developed for that purpose. In this regard, the trunk is considered to be a 3D geometric shape, and a clustering method sharing the following features was developed: (1) the Euclidean distance between the 3D points must be lower than 50 cm (this value is estimated considering the typical distance between grapevines, which is never of this order); and (2) the minimum number of points for clustering is set to five. According to these constraints, a correct limitation of the growing region was determined for each cluster. Therefore, k groups of points were segmented for each vine row.

Regarding the results of spatial clustering, two optimisations are considered to improve the final results. The first optimisation focuses on solving errors related to the spatial segmentation. In areas characterised by dense vegetation, where the trunk was partially occluded by leaves, the trunk's 3D reconstruction is, in general, generated with a lower detail. In these cases, the trunk is potentially composed by a few real points and many noisy points. Consequently, an inaccurate segmentation is achieved, which causes false positives in those regions. This issue is overtaken by testing the angle (α) between two vectors: (1) the direction vector of the vine row axis; (2) the direction formed by two consecutive centroids of clusters. A maximum deviation of 20° is allowed. This value proved to be adequate to remove clusters which are not correctly segmented. As illustrated in Figure 8, green points are valid centroids and red points are those centroids which are discarded. The blue arrows depict the correct patch formed by all plants of the row. This test is performed along the vine row in the same direction on the x-axis.

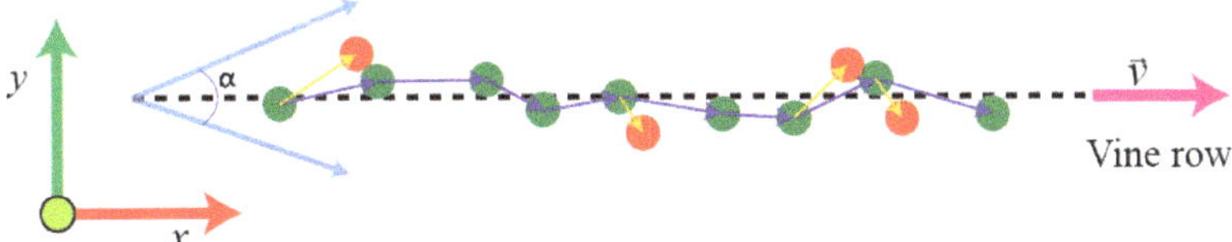

Figure 8. Optimisation of the clustering segmentation procedure.

The second optimisation consists on the automatic recognition of posts, which are considered to be trunks, since geometrically both are very similar. For this purpose, key geometric features were considered to detect posts for each vine row. In Figure 9a, trunks and posts are marked by a green and yellow rectangle, respectively. The main difference is the number of vegetation points around them. As shown in Figure 9b, a neighbour search is developed for each centroid, limited by a radius (r = 80 cm). This value is the most adequate considering the mean size of grapevines in the dataset. To check points which were into this spherical-searching region, Equation (4) is applied. Then, the point colour is also considered in order to detect vegetation points. Vegetation points are characterised by a threshold for the R, G and B channels so, if some of them are located inside the search area, the cluster is considered to be trunk. Likewise, if no vegetation points are found, the cluster is classified as a post.

$$x^2 + y^2 + z^2 - r^2 < 0 \tag{4}$$

where for each cluster, x, y and z represent the coordinates of a 3D point and r the radius of the search region.

Figure 9. Recognition of posts: (**a**) post location in the point cloud; and (**b**) the search of vegetation points around the trunk.

2.3.5. Estimation of Missing Plants and Occluded Trunks

Depending on the epoch of the year in which data are acquired, vegetation may occlude trunks (Figure 2). The proposed method is also prepared to deal with such scenarios, being able to estimate the position of no visible trunks and missing plants. In general, and mostly in recent commercial vineyards, each vine row is formed by plants which are equidistant from each other. However, the presented method estimates the space between plants automatically, which makes its use universal and fully automatic. For testing this feature, the method was applied on a complex vineyard plot presented in Figure 1. Therefore, the method can be used in any vineyard even in those which present challenging features such as the irregular distance between plants, replanted grapevines with different trunk diameter and some plants not visible from aerial images due to a dense foliage (Figure 10).

Figure 10. Cases where the trunks are occluded or cannot properly modelled.

The estimation of missing plants and occluded trunks is useful in order to optimise the results of plant recognition as well as to know the number of voids along the vine rows, which could be occupied by new plants. For this purpose, the resulting data from the individual grapevine detection were used in order to identify areas where no plants were found. Firstly, the distances between consecutive grapevines are calculated and sorted from the lowest to the highest, considering all plants detected in a given vine row.

Then, all distances less than 50 cm are discarded and the top 10% values of the sorted list are used to calculate the average distance (*d*). This value was calculated for each vine row and is also used to highlight areas that can potentially contain missing plants or occluded grapevines. If the distance between two consecutive plants (*D*) is higher than *d*, and the rate d/D is higher than one, the integer part of the quotient represents the number of plants that should be detected in the area. Then, missing plants and/or occluded trunks are marked using the average distance (*d*) as reference. Figure 11 illustrates the representation of two detected grapevines and the area between them where there are missing plants and occluded trunks. The last step consists in determining if the marked point represents a missing plant or an occluded trunk. To this end, a point inclusion test in a three-dimensional cylinder (height of 1 m and a radius of 20 cm) is implemented, considering the same constrains which were used for the trunk detection.

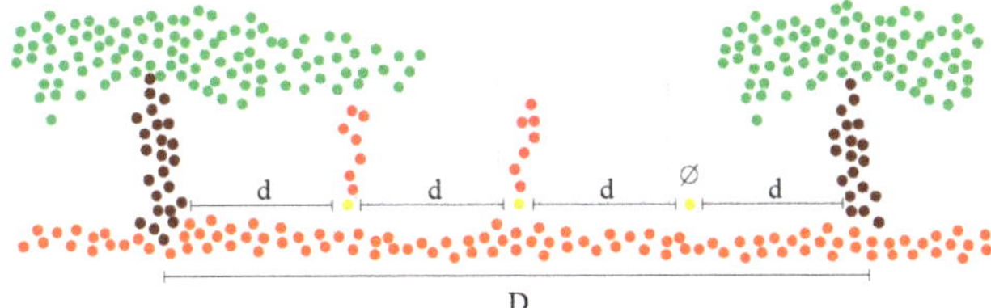

Figure 11. Recognition of missing or occluded plants.

2.4. Validation Process

To analyse the robustness and effectiveness of the proposed method in all its features, five vine rows of the complex study area (polygon highlighted in Figure 1) were used. A field campaign was performed in order to map the real state of vine rows and to determine the location of missing plants. This way, for each vine row, the results provided by the application of the proposed method were compared with ground-truth data allowing computing the overall accuracy of the whole validation area and of each estimation.

Grapevine estimation evaluation was conducted based on the number of correct (true positive—TP) and incorrect (false positive—FP) grapevine estimations and also considering the correct/incorrect estimation of missing plants along the vine rows as, respectively, true negatives (TN) and false negatives (FN). From theses data precision, recall, F1score and the overall accuracy were computed for each vine row and for the whole validation area.

3. Results

3.1. Point Cloud Reconstruction and Processing

The proposed method works perfectly when it is applied to well-maintained commercial vineyards plots, using aerial data acquired as described in Section 2.2. In those cases, the method is able to detect all existing trunks. However, the method was pushed to its limit, being applied to a complex area, arising from the existence of distinct vegetation density/vigour areas, several voids caused by missing plants, and new plants that were replaced defective or dead plants.

As results, a 3D point cloud formed by 26,656,371 points was generated and the time for point cloud densification was 01 h:29 m:23 s. Figure 12 shows the generated point cloud and the virtual lines that represent the vine rows axis. Most of plants could be fully modelled, but there are some grapevines whose trunks were partially occluded by leaves. Moreover, noisy points were produced around the trunks, between the leaves of plants and the ground. These 3D points negatively affect the recognition of trunk's shape. To address this problem, a noise filter was applied and most of noising points could be removed. This process was applied using a kernel size of 0.05 m. In summary, 10.448.046 points were discarded in the generated dense point cloud.

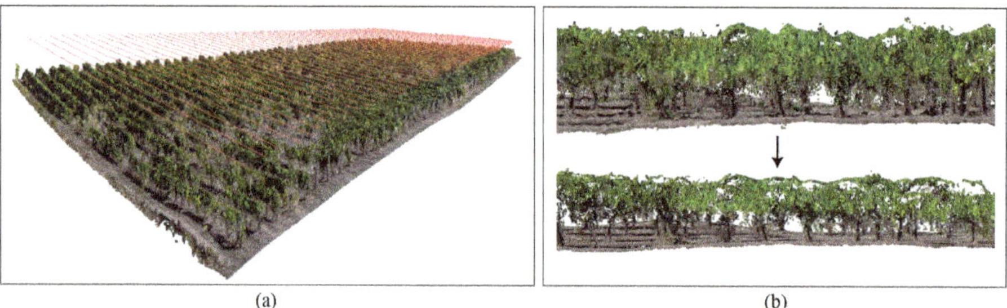

Figure 12. 3D model generated of the complex vineyard plot: (**a**) the reconstruction of study area using all the 3D points; and (**b**) final model, after application of noise filter.

3.2. *Individual Grapevine Detection*

Once the point cloud is generated and the noisy points filtered out, the 3D model is segmented in order to discard ground and leaves points. Figure 13 shows the results of this step, when the method is applied to the vine rows of the validation area. Consequently, red points (classified as ground and leaves, Figure 13b) are discarded and just trunk points, which will be used as input data for the spatial clustering, remain.

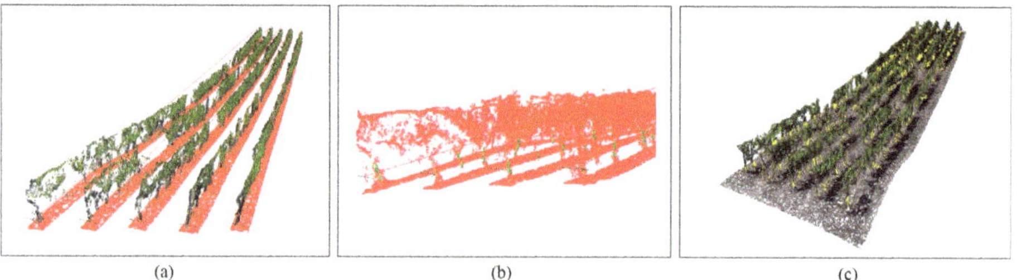

Figure 13. Main steps for individual trunk detection: (**a**) ground points identification and removal, (**b**) vegetation/leaf points identification and removal; and (**c**) trunk detection.

The method was applied to the whole plot and the location of detected and missing plants was analysed in the QGIS software. This output is presented in Figure 14, with the orthophoto mosaic in the background. A total of 1916 grapevines were estimated and 402 plants were classified as being missing.

Regarding the efficiency of the method, the time required for the automatic recognition of each individual grapevine in the whole plantation was 38 s using a PC with CPU (Intel Xeon(R) W-2145) and RAM (64 GB). The low time required for computing the methods makes possible the use of portable devices for on-site processing.

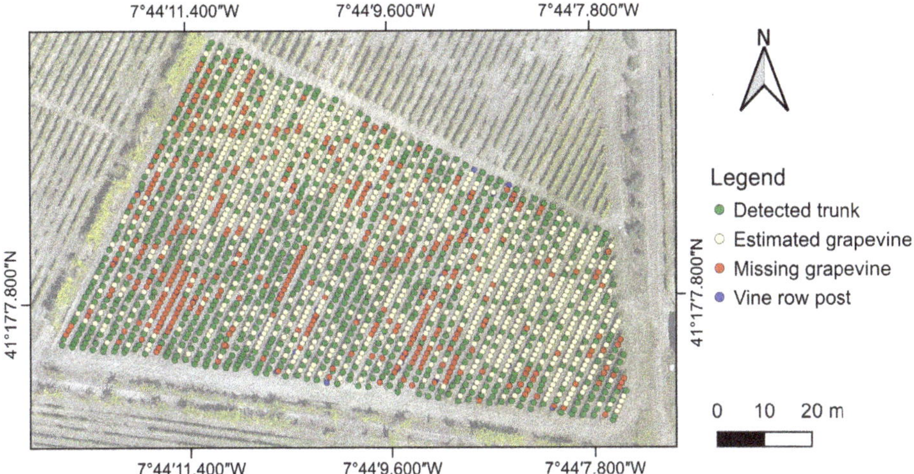

Figure 14. General overview of the results obtained from the application of the proposed method to the whole complex plot. Coordinates in WGS84 (EPSG:4326).

3.3. Grapevine Estimation Accuracy

The results of the grapevine estimation from the application of the proposed method to the five validation vine rows are presented in Table 1.

Table 1. Number of plants and overall accuracy (OA) of the proposed method compared to the ground-truth data of five vine rows. Row total values—maximum possible number of plants in a given vine row—are also provided.

Vine Row	Number of Grapevines			Missing Grapevines			Row Total		
	Obs.	Est.	OA (%)	Obs.	Est.	OA (%)	Obs.	Est.	OA (%)
1	46	43	93.5	12	15	75.0	58	58	100.0
2	39	37	94.9	18	21	83.3	57	58	98.2
3	42	45	92.9	15	12	80.0	57	57	100.0
4	40	46	85.0	17	9	52.9	57	55	96.5
5	49	50	98.0	7	7	100.0	56	57	98.2
Total	216	221	97.7	69	64	92.8	285	285	100.0

Regarding the total number of grapevines presented in the evaluated vine rows (ranging between 39 to 46), an overestimation of five plants is observed. Under detection of grapevines is verified in two rows, differing in three and two grapevines, respectively. The opposite is observed in the other three vine rows, with ten plants being overestimated (respectively, three, six and one). According to the number of missing grapevines (69, in total, ranging between 7 to 18 per row, average of 14 missing plants) the results obtained by the method show an overestimation of six missing grapevines in two vine rows (three in each) and one vine row is in agreement with the ground-truth data. As for the other vine rows the under estimation of missing plants diverges from three to eight plants (64 missing plants in total, with an average of 13 missing plants per row, 93% overall accuracy). As for the total number of possible grapevines in a given vine row (sum of the grapevines and missing grapevines), a mean of 57 plants were estimated, the same number when observing the ground-truth data, being the total also 285 plants, obtaining three rows with the same number of plants, two rows with one plant less, and one row with two plants more. For this specific parameter, the overall accuracy ranges between 97% (underestimation) and 100% considering all vine rows. According to the capability of the proposed method in the automatic detection of grapevines, 157 grapevines (71%) were directly detected from the grapevine trunk, and by the analysis of the point cloud density 64 grapevines were estimated. Moreover, several posts were also detected along the vine rows and therefore, were automatically discarded.

To further validate the spatial accuracy of the methods outputs, each detection was analysed to assess if the estimations were correctly located. By using the ground-truth data, false negatives and false positives results—grapevines classified as being missing and the inverse, missing plants classified as being grapevines—are evaluated. Table 2 presents these results. In total, from the 221 estimated plants 197 plants were correctly detected and classified (89%), as for missing plants, 41 (64%) were correctly classified, the overall accuracy considering all data is approximately 84%.

Table 2. Evaluation of the proposed method in the classification grapevines and missing grapevines for the following parameters: precision, recall, F1score, and overall accuracy (OA). The ground-truth of five vine rows was used. TP: true positive; FP: false positive; TN: true negative; FN: false negative.

Vine Row	TP	FP	TN	FN	Precision	Recall	F1score	O.A. (%)
1	39	4	8	7	0.91	0.85	0.88	81.0
2	32	5	16	5	0.86	0.86	0.86	82.8
3	40	5	7	5	0.89	0.89	0.89	82.5
4	42	4	6	3	0.91	0.93	0.92	87.3
5	45	5	4	3	0.90	0.94	0.92	86.0
Total	198	23	41	23	0.90	0.90	0.90	83.9

Figure 15 shows the 3D location of each plant on the test area. Green points represent the centroids of plants that are directly detected in the 3D model. The centroids' position were then checked against its corresponding vine row in order to identify centroids wrongly classified as trunk (red points). These wrong classifications are caused due to points of vegetation, which are around the trunk and could not be totally removed by the noise filter. Then, posts were distinguished from the grapevine trunks (blue points). Finally, missing plants (pink points) and occluded trunks (yellow points) were estimated. In terms of quantitative data, in the validation area, the proposed method was able to detect 221 plants and 64 missing plants.

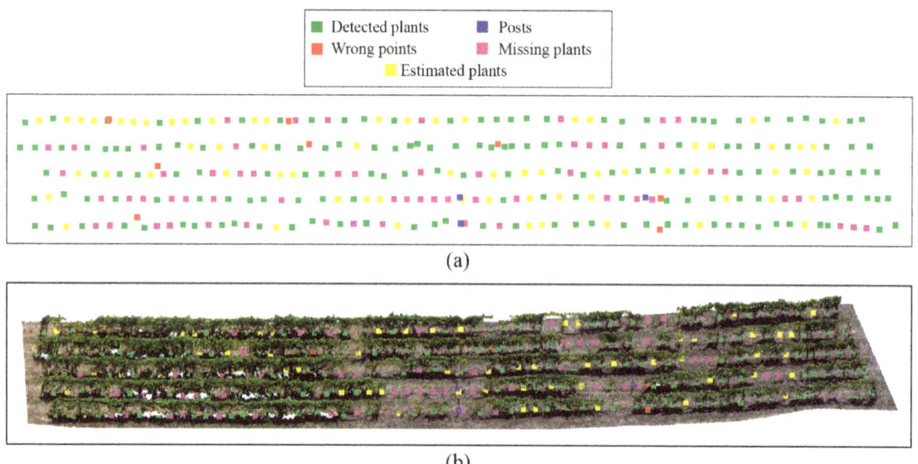

Figure 15. Individual grapevine delineation resulting from the application of the proposed method to the validation area: (**a**) points to represent detected plants (visible and occluded trunks), missing plants, wrong clusters and posts; and (**b**) 3D point cloud and points computed for the plant location.

4. Discussion

4.1. Point Cloud Reconstruction and Processing

The generation of point clouds for remote sensing applications was enhanced by the proliferation of innovative UAV-based technologies such as high-resolution cameras and LiDAR systems. By applying photogrammetric techniques, point clouds can be obtained using multiple overlapping images. Other option is the use of LiDAR scanners which provide dense point clouds of natural environments but these are more expensive than digital cameras [34]. LiDAR data discriminate better plant's canopy since it penetrates vegetation [35,36], and, therefore, potentially making the identification of the grapevine trunks easier. Moreover, photogrammetric techniques tend to estimate erroneous points in the cases where some points from the ground are estimated along a post. One of the advantages of the proposed method is that it remains operational even when using point cloud data from other type of sensors, reinforcing that when LiDAR sensors for UAVs become more affordable the method can still be employed. The proposed method is not dependent on any technology, although to get proper results is required enough geometric quality of the 3D model to ensure a partial reconstruction of the trunks at least. In this sense, to avoid noisy points around the trunk and vegetation, the presented solution integrates a noise filter. Hence, the proposed method can be applied using any point cloud data with accurate results.

4.2. Individual Grapevine Detection

There are several methods for the automatic detection and parameters extraction on 3D models using crop height models [37], combining terrestrial laser scanner and UAV photogrammetric point clouds [38], fusing RGB and multispectral point clouds to extract individual tree parameters [39], computing 3D vegetation indices in olive groves [40], and obtaining forest structural attributes [41]. However, given the complexity and the unique characteristics of vineyard plots, such methods are not suitable to be applied. Studies focusing on the use of photogrammetric point clouds, generated from UAV-based imagery, were dealing with vineyard detection [23] or detect and describe some of its general properties [17,25,42]. Studies for individual grapevine detection using UAV-based raster outcomes often rely on the coarse position of each vine assuming a mean distance of separation between grapevines along the vine row [13,20,21]. The proposed method addresses all these limitations, using point cloud data and geometrical characteristics, to automatically identify points belonging to grapevine's trunk. Hence, individual plants can be detected and missing plants estimated.

The complex vineyard plot analysed in this study had been used in another study [13] with data acquired in 2018. The whole plot contained 2266 plants and was evaluated by the method with an accuracy of 98%. However, the aerial imagery was acquired in an early phase of the vegetative state [43] with a lower vegetation density, which can help in the detection of missing plants. Furthermore, a constant distance between individual plant was also used. The results presented in Section 2.4 present an overestimation of 2% (Table 1) according to the number of grapevines and an under-estimation of 8% considering the number of missing plants. This fact is related to the presence of vegetation from adjacent grapevines in areas with missing plants that could cover the void. The results are improved by selecting an early period (preferably belonging to phase 2—Figure 2) to conduct grapevine detection, with grapevines with a lower leaf cover, preventing the existence of vegetation in areas with no plants [21]. In Di Gennaro and Matese [44], 3D and 2.5D methods were compared for vineyard biomass and plant detection, but individual grapevine detection was not performed. In that study, as for missing plants, an overestimation was observed in the 3D-based method, false negatives were related to the existence of new plants while some false positives were due to different grapevine canopy thickness. Similar findings were detected in the miss-classifications observed in this study (Table 2). Moreover, proximal sensing approaches for grapevine trunk detection using ground vehicles were also tested by research groups, either using LiDAR [45,46] or depth cameras [47]. However, in contrast to UAVs, such approaches are more expensive—due to the equipment used—and time-consuming, since the vehicles need to go through all vine rows, where some obstacles can also be present in their way.

5. Conclusions

The innovative method presented in this study is proved to be effective for a rapid access of the vineyard status using UAV-based 3D point clouds, with automation levels that allow its applicability in different vineyards, not relying on predefined parameters as the distance among plants. The proposed method is able to detect occluded trunks with reliable accuracy rates and missing plants, where in the vineyard context represents the occurrence of voids along the vine rows. The major contribution of this work is that the approach is fully automatic, not requiring any a prior knowledge of the distance between plants or number of plants per row, as in existing approaches. Moreover, the computational complexity of the proposed technique does not require high-performance computing and is appropriate for use on mobile in-field computing devices.

The applicability of the proposed method can be extended to other types of purposes related to the estimation of biophysical parameters of grapevines, providing a more efficient understanding of data for vineyard management and the validation of the use of UAV-based point clouds. Indeed, its impact is increased in a multi-temporal context. In this way, the estimated canopy of each detected grapevine can be studied to measure its volume which can help in the decision-making process for canopy management operations and, consequently, yield optimisation. To improve data quality and to extend the method capabilities, an in-depth investigation of the flight parameters optimisation (flight height, imagery overlap, camera angle) is required, which can also make possible an automatic detection grape bunches. The results of this study might influence on further research related to individual monitoring of every grapevine, multi-temporal studies and making accurate support decision systems for an optimal vineyard management.

Author Contributions: Conceptualisation, J.M.J., L.P. and J.J.S.; methodology, J.M.J., L.P. and J.J.S.; software, J.M.J.; validation, F.R.F. and J.J.S.; formal analysis, L.P.; investigation, J.M.J. and L.P.; resources, F.R.F. and J.J.S.; data curation, J.M.J. and L.P.; writing—original draft preparation, J.M.J. and L.P.; writing—review and editing, F.R.F. and J.J.S.; visualisation, J.M.J. and L.P.; supervision, F.R.F. and J.J.S.; project administration, J.J.S.; funding acquisition, F.R.F. and J.J.S. All authors have read and agreed to the published version of the manuscript.

Funding: This research was partially funded by the Ministry of Science and Innovation of Spain and the European Union (via ERDF funds), through the research project TIN2017-84968-R, by the EDUJA grant of University of Jaén to Juan M. Jurado, and by the FCT-Portuguese Foundation for Science and Technology (SFRH/BD/139702/2018) to Luís Pádua.

Conflicts of Interest: The authors declare no conflict of interest.

References

1. Zhang, C.; Kovacs, J.M. The application of small unmanned aerial systems for precision agriculture: A review. *Precis. Agric.* **2012**, *13*, 693–712. [CrossRef]
2. Mogili, U.R.; Deepak, B. Review on application of drone systems in precision agriculture. *Procedia Comput. Sci.* **2018**, *133*, 502–509. [CrossRef]
3. Bernardes, M.F.F.; Pazin, M.; Pereira, L.C.; Dorta, D.J. Impact of pesticides on environmental and human health. *Toxicol. Stud. Cells Drugs Environ.* **2015**, 195–233.
4. Scott, G.; Rajabifard, A. Sustainable development and geospatial information: A strategic framework for integrating a global policy agenda into national geospatial capabilities. *Geo-Spat. Inf. Sci.* **2017**, *20*, 59–76. [CrossRef]
5. Pádua, L.; Vanko, J.; Hruška, J.; Adão, T.; Sousa, J.J.; Peres, E.; Morais, R. UAS, sensors, and data processing in agroforestry: A review towards practical applications. *Int. J. Remote Sens.* **2017**, *38*, 2349–2391. [CrossRef]
6. Ezenne, G.; Jupp, L.; Mantel, S.; Tanner, J. Current and potential capabilities of UAS for crop water productivity in precision agriculture. *Agric. Water Manag.* **2019**, *218*, 158–164. [CrossRef]
7. Shi, X.; Han, W.; Zhao, T.; Tang, J. Decision support system for variable rate irrigation based on UAV multispectral remote sensing. *Sensors* **2019**, *19*, 2880. [CrossRef] [PubMed]

8. Zhang, M.; Zhou, J.; Sudduth, K.A.; Kitchen, N.R. Estimation of maize yield and effects of variable-rate nitrogen application using UAV-based RGB imagery. *Biosyst. Eng.* **2020**, *189*, 24–35. [CrossRef]
9. Mendes, J.; Pinho, T.M.; Neves dos Santos, F.; Sousa, J.J.; Peres, E.; Boaventura-Cunha, J.; Cunha, M.; Morais, R. Smartphone Applications Targeting Precision Agriculture Practices—A Systematic Review. *Agronomy* **2020**, *10*, 855. [CrossRef]
10. Matese, A.; Di Gennaro, S.F. Technology in precision viticulture: A state of the art review. *Int. J. Wine Res.* **2015**, *7*, 69–81. [CrossRef]
11. Proffitt, A.P.B.; Bramley, R.; Lamb, D.; Winter, E. *Precision Viticulture: A New Era in Vineyard Management and Wine Production*; Winetitles Pty Ltd.: Ashford, SA, USA, 2006.
12. Campos, J.; Llop, J.; Gallart, M.; García-Ruiz, F.; Gras, A.; Salcedo, R.; Gil, E. Development of canopy vigour maps using UAV for site-specific management during vineyard spraying process. *Precis. Agric.* **2019**, *20*, 1136–1156. [CrossRef]
13. Pádua, L.; Adão, T.; Sousa, A.; Peres, E.; Sousa, J.J. Individual Grapevine Analysis in a Multi-Temporal Context Using UAV-Based Multi-Sensor Imagery. *Remote Sens.* **2020**, *12*, 139. [CrossRef]
14. Pádua, L.; Marques, P.; Hruška, J.; Adão, T.; Bessa, J.; Sousa, A.; Peres, E.; Morais, R.; Sousa, J.J. Vineyard properties extraction combining UAS-based RGB imagery with elevation data. *Int. J. Remote Sens.* **2018**, *39*, 5377–5401. [CrossRef]
15. Comba, L.; Gay, P.; Primicerio, J.; Aimonino, D.R. Vineyard detection from unmanned aerial systems images. *Comput. Electron. Agric.* **2015**, *114*, 78–87. [CrossRef]
16. Mathews, A.J.; Jensen, J.L. Visualizing and quantifying vineyard canopy LAI using an unmanned aerial vehicle (UAV) collected high density structure from motion point cloud. *Remote Sens.* **2013**, *5*, 2164–2183. [CrossRef]
17. Weiss, M.; Baret, F. Using 3D point clouds derived from UAV RGB imagery to describe vineyard 3D macro-structure. *Remote Sens.* **2017**, *9*, 111. [CrossRef]
18. Poblete-Echeverría, C.; Olmedo, G.F.; Ingram, B.; Bardeen, M. Detection and segmentation of vine canopy in ultra-high spatial resolution RGB imagery obtained from unmanned aerial vehicle (UAV): A case study in a commercial vineyard. *Remote Sens.* **2017**, *9*, 268. [CrossRef]
19. Caruso, G.; Tozzini, L.; Rallo, G.; Primicerio, J.; Moriondo, M.; Palai, G.; Gucci, R. Estimating biophysical and geometrical parameters of grapevine canopies ('Sangiovese') by an unmanned aerial vehicle (UAV) and VIS-NIR cameras. *Vitis* **2017**, *56*, 63–70.
20. De Castro, A.I.; Jimenez-Brenes, F.M.; Torres-Sánchez, J.; Peña, J.M.; Borra-Serrano, I.; López-Granados, F. 3-D characterization of vineyards using a novel UAV imagery-based OBIA procedure for precision viticulture applications. *Remote Sens.* **2018**, *10*, 584. [CrossRef]
21. Matese, A.; Di Gennaro, S.F. Practical applications of a multisensor uav platform based on multispectral, thermal and rgb high resolution images in precision viticulture. *Agriculture* **2018**, *8*, 116. [CrossRef]
22. Primicerio, J.; Caruso, G.; Comba, L.; Crisci, A.; Gay, P.; Guidoni, S.; Genesio, L.; Ricauda Aimonino, D.; Vaccari, F.P. Individual plant definition and missing plant characterization in vineyards from high-resolution UAV imagery. *Eur. J. Remote Sens.* **2017**, *50*, 179–186. [CrossRef]
23. Comba, L.; Biglia, A.; Aimonino, D.R.; Gay, P. Unsupervised detection of vineyards by 3D point-cloud UAV photogrammetry for precision agriculture. *Comput. Electron. Agric.* **2018**, *155*, 84–95. [CrossRef]
24. Comba, L.; Biglia, A.; Aimonino, D.R.; Tortia, C.; Mania, E.; Guidoni, S.; Gay, P. Leaf Area Index evaluation in vineyards using 3D point clouds from UAV imagery. *Precis. Agric.* **2020**, *21*, 881–896. [CrossRef]
25. Mesas-Carrascosa, F.J.; de Castro, A.I.; Torres-Sánchez, J.; Triviño-Tarradas, P.; Jiménez-Brenes, F.M.; García-Ferrer, A.; López-Granados, F. Classification of 3D point clouds using color vegetation indices for precision viticulture and digitizing applications. *Remote Sens.* **2020**, *12*, 317. [CrossRef]
26. Aboutalebi, M.; Torres-Rua, A.F.; McKee, M.; Kustas, W.P.; Nieto, H.; Alsina, M.M.; White, A.; Prueger, J.H.; McKee, L.; Alfieri, J.; et al. Incorporation of Unmanned Aerial Vehicle (UAV) Point Cloud Products into Remote Sensing Evapotranspiration Models. *Remote Sens.* **2020**, *12*, 50. [CrossRef]
27. Moreno, H.; Valero, C.; Bengochea-Guevara, J.M.; Ribeiro, Á.; Garrido-Izard, M.; Andújar, D. On-Ground Vineyard Reconstruction Using a LiDAR-Based Automated System. *Sensors* **2020**, *20*, 1102. [CrossRef]
28. Rusu, R.B.; Cousins, S. 3d is here: Point cloud library (pcl). In Proceedings of the 2011 IEEE International Conference on Robotics and Automation, Shanghai, China, 9–13 May 2011; IEEE: Piscataway, NJ, USA, 2011; pp. 1–4.

29. Schönberger, J.L.; Frahm, J. Structure-from-Motion Revisited. In Proceedings of the 2016 IEEE Conference on Computer Vision and Pattern Recognition (CVPR), Las Vegas, NV, USA, 26 June–1 July 2016; pp. 4104–4113. [CrossRef]
30. Colomina, I.; Molina, P. Unmanned aerial systems for photogrammetry and remote sensing: A review. *ISPRS J. Photogramm. Remote Sens.* **2014**, *92*, 79–97. [CrossRef]
31. Han, X.F.; Jin, J.S.; Wang, M.J.; Jiang, W.; Gao, L.; Xiao, L. A review of algorithms for filtering the 3D point cloud. *Signal Process. Image Commun.* **2017**, *57*, 103–112. [CrossRef]
32. Richardson, A.D.; Jenkins, J.P.; Braswell, B.H.; Hollinger, D.Y.; Ollinger, S.V.; Smith, M.L. Use of digital webcam images to track spring green-up in a deciduous broadleaf forest. *Oecologia* **2007**, *152*, 323–334. [CrossRef]
33. Otsu, N. A threshold selection method from gray-level histograms. *IEEE Trans. Syst. Man Cybern.* **1979**, *9*, 62–66. [CrossRef]
34. Chen, S.; McDermid, G.J.; Castilla, G.; Linke, J. Measuring vegetation height in linear disturbances in the boreal forest with UAV photogrammetry. *Remote Sens.* **2017**, *9*, 1257. [CrossRef]
35. Lisein, J.; Pierrot-Deseilligny, M.; Bonnet, S.; Lejeune, P. A photogrammetric workflow for the creation of a forest canopy height model from small unmanned aerial system imagery. *Forests* **2013**, *4*, 922–944. [CrossRef]
36. Guimarães, N.; Pádua, L.; Marques, P.; Silva, N.; Peres, E.; Sousa, J.J. Forestry Remote Sensing from Unmanned Aerial Vehicles: A Review Focusing on the Data, Processing and Potentialities. *Remote Sens.* **2020**, *12*, 1046. [CrossRef]
37. Panagiotidis, D.; Abdollahnejad, A.; Surový, P.; Chiteculo, V. Determining tree height and crown diameter from high-resolution UAV imagery. *Int. J. Remote Sens.* **2017**, *38*, 2392–2410. [CrossRef]
38. Tian, J.; Dai, T.; Li, H.; Liao, C.; Teng, W.; Hu, Q.; Ma, W.; Xu, Y. A novel tree height extraction approach for individual trees by combining TLS and UAV image-based point cloud integration. *Forests* **2019**, *10*, 537. [CrossRef]
39. Jurado, J.M.; Ramos, M.; Enríquez, C.; Feito, F. The Impact of Canopy Reflectance on the 3D Structure of Individual Trees in a Mediterranean Forest. *Remote Sens.* **2020**, *12*, 1430. [CrossRef]
40. Jurado, J.M.; Ortega, L.; Cubillas, J.J.; Feito, F. Multispectral mapping on 3D models and multi-temporal monitoring for individual characterization of olive trees. *Remote Sens.* **2020**, *12*, 1106. [CrossRef]
41. Cao, L.; Liu, H.; Fu, X.; Zhang, Z.; Shen, X.; Ruan, H. Comparison of UAV LiDAR and digital aerial photogrammetry point clouds for estimating forest structural attributes in subtropical planted forests. *Forests* **2019**, *10*, 145. [CrossRef]
42. Comba, L.; Zaman, S.; Biglia, A.; Ricauda, A.D.; Dabbene, F.; Gay, P. Semantic interpretation and complexity reduction of 3D point clouds of vineyards. *Biosyst. Eng.* **2020**, *197*, 216–230. [CrossRef]
43. Magalhães, N. *Tratado de Viticultura: A Videira, a Vinha e o Terroir*; Publicações Chaves Ferreira Lisboa: Lisboa, Portugal , 2008; 605p, ISBN 9789899820739.
44. Di Gennaro, S.F.; Matese, A. Evaluation of novel precision viticulture tool for canopy biomass estimation and missing plant detection based on 2.5 D and 3D approaches using RGB images acquired by UAV platform. *Plant Methods* **2020**, *16*, 1–12. [CrossRef]
45. Siebers, M.H.; Edwards, E.J.; Jimenez-Berni, J.A.; Thomas, M.R.; Salim, M.; Walker, R.R. Fast phenomics in vineyards: Development of GRover, the grapevine rover, and LiDAR for assessing grapevine traits in the field. *Sensors* **2018**, *18*, 2924. [CrossRef] [PubMed]
46. Milella, A.; Marani, R.; Petitti, A.; Reina, G. In-field high throughput grapevine phenotyping with a consumer-grade depth camera. *Comput. Electron. Agric.* **2019**, *156*, 293–306. [CrossRef]
47. Mendes, J.; Dos Santos, F.N.; Ferraz, N.; Couto, P.; Morais, R. Vine trunk detector for a reliable robot localization system. In Proceedings of the 2016 International Conference on Autonomous Robot Systems and Competitions (ICARSC), Bragança, Portugal, 4–6 May 2016; IEEE: Piscataway, NJ, USA, 2016; pp. 1–6.

Article

UAV-Borne LiDAR Crop Point Cloud Enhancement Using Grasshopper Optimization and Point Cloud Up-Sampling Network

Jian Chen [1], Zichao Zhang [1], Kai Zhang [1], Shubo Wang [1] and Yu Han [2,*]

1 College of Engineering, China Agricultural University, Beijing 100083, China; jchen@cau.edu.cn (J.C.); zhangzc1@cau.edu.cn (Z.Z.); sy20193071234@cau.edu.cn (K.Z.); wangshubo@cau.edu.cn (S.W.)

2 College of Water Resources & Civil Engineering, China Agricultural University, Beijing 100083, China

* Correspondence: yhan@cau.edu.cn

Received: 5 September 2020; Accepted: 28 September 2020; Published: 1 October 2020

Abstract: Because of low accuracy and density of crop point clouds obtained by the Unmanned Aerial Vehicle (UAV)-borne Light Detection and Ranging (LiDAR) scanning system of UAV, an integrated navigation and positioning optimization method based on the grasshopper optimization algorithm (GOA) and a point cloud density enhancement method were proposed. Firstly, a global positioning system (GPS)/inertial navigation system (INS) integrated navigation and positioning information fusion method based on a Kalman filter was constructed. Then, the GOA was employed to find the optimal solution by iterating the system noise variance matrix Q and measurement noise variance matrix R of Kalman filter. By feeding the optimal solution into the Kalman filter, the error variances of longitude were reduced to 0.00046 from 0.0091, and the error variances of latitude were reduced to 0.00034 from 0.0047. Based on the integrated navigation, an UAV-borne LiDAR scanning system was built for obtaining the crop point. During offline processing, the crop point cloud was filtered and transformed into WGS-84, the density clustering algorithm improved by the particle swarm optimization (PSO) algorithm was employed to the clustering segment. After the clustering segment, the pre-trained Point Cloud Up-Sampling Network (PU-net) was used for density enhancement of point cloud data and to carry out three-dimensional reconstruction. The features of the crop point cloud were kept under the processing of reconstruction model; meanwhile, the density of the crop point cloud was quadrupled.

Keywords: UAV-borne LiDAR scanning system; grasshopper optimization algorithm; GPS/INS integrated navigation; point cloud up-sampling network (PU-net); clustering segmentation; 3-dimensional reconstruction

1. Introduction

Light Detection and Ranging (LiDAR) is an active sensing technology that can quickly acquire spatial information of a target or environment [1,2]. Compared with traditional remote sensing methods, LiDAR has shown great advantages in accuracy and stability [3]. In agriculture, the LiDAR system is one of the most accurate methods to measure regional vegetation structural characteristics and biophysical parameters [4]. The Unmanned Aerial Vehicle (UAV)-borne LiDAR scanning system is one of the most common platforms for crop point cloud obtaining. The crop point clouds captured by UAV-borne LiDAR scanning system are sparse and unordered [5]. The navigation and positioning accuracy of the UAV-borne LiDAR scanning system influences the accuracy of the point cloud data greatly. Density is also an important indicator to measure the quality of point cloud data. Higher point cloud density represents richer information of the target or environment [6]. Ensuring the accuracy and density characteristics of the UAV-borne LiDAR point cloud data while maintaining the cost is crucial for the continuous and efficient execution of LiDAR detection tasks.

For integrated navigation, the most widely used in the world is the global positioning system (GPS)/inertial navigation system (INS) integrated navigation positioning system [7]. The Kalman filter is the most common method of data fusion [8]. Gao et al. [9] developed an INS/GPS/LiDAR integrated navigation system through an extended Kalman filter (EKF) using a hybrid scan matching algorithm. Shabani et al. [10] explored the characteristics, advantages, and disadvantages of direct Kalman filtering and indirect Kalman filtering commonly used in GPS/INS integrated navigation, and a novel asynchronous direct Kalman filter was proposed. Zhang et al. [11] developed a low-cost global navigation satellite system (GNSS) and INS integrated navigation method and an adaptive Kalman filter for overcoming the measurement noise of GNSS, which was proposed based on a supervised machine learning model. Yu et al. [12] used a single-frequency real-time kinematic carrier phase (RTK) technology to deduce and obtain information including speed and orientation under GPS/BeiDou Navigation Satellite System (BDS) for polishing the inaccurate INS information. Norouz et al. [13] proposed an improved Kalman filter algorithm based on small probability distribution of the state variable between the measurement updates of the multi-rate GPS/INS-coupled navigation filter, which greatly reduced the amount of calculation, and the algorithm performed well by numerically simulating. All of the above integrated navigation systems suffered from the low accuracy of integrated navigation, except Yu et al. [12]. However, RTK-based high accuracy integrated navigation systems cannot be widely popularized due to the price gap. During the experiments of data fusion using the Kalman filter, the longitude and latitude errors were sharply affected by the parameter of the Kalman filter, especially system noise sequence Q and measurement noise sequence R, and the same situation also occurred in the realization of Ref. [10,11]. For promoting the accuracy of our integrated navigation-based scanning system, the meta-heuristic method was employed for finding the optimum parameters of integrated navigation in our UAV scanning system. For multi-solution problems, genetic algorithms (GA) [14], particle swarm optimization (PSO) [15], and ant colony optimization (ACO) [16] were the most popular, with swarm intelligence optimization proposed recently such as the bat algorithm (BA) [17] and grasshopper optimization algorithm (GOA) [18]. All of the multi-solution meta-heuristic methods have their own merits for different problems [19]. In order to reduce longitude and latitude errors of integrated navigation, the meta-heuristic methods were employed to find the optimum system noise sequence Q and measurement noise sequence R.

After obtaining the crop point cloud data by UAV scanning system, the raw point cloud data consisted of crop points and other points, and manual classification was extremely inefficient for different point cloud data. Therefore, the clustering segmentation method was employed to extract crop data from raw point cloud data. The density-based clustering method, density-based spatial clustering of applications with noise (DBSCAN), has attracted a lot of attention; besides the improvement of the algorithm itself, finding out the key parameters of clustering was also a new direction of optimization [20]. The same as integrated navigation, the meta-heuristic method was employed to optimize the clustering segmentation method. According to the experimental results of the density distribution characteristics of the plane point cloud under different conditions, Kedzierski et al. [21] found that the point cloud density could be preliminarily estimated based on distance and angle, but the actual point cloud density should take into account the influence of object shape and environment. Rupink et al. [22] proposed the formula and calculation steps for calculating the point cloud density. Huang et al. [23] proposed an edge-sensing point set resampling method, gradually approaching the edges and corners. However, the quality of the results largely depends on the accuracy of the normal of a given point and the fine-tuning of parameters. Based on Ref. [21,23], dense and accurate point clouds are very useful for crop point cloud recognition and 3D reconstruction. The high-cost multi-line LiDAR can be simulated by low-cost single-line LiDAR point clouds through the point cloud up-sampling method without destroying the features.

In view of the current problems and inspired by existing methods, in order to improve the quality of point cloud data, a combined navigation optimization algorithm based on GOA and a point cloud density enhancement algorithm based on Point Cloud Up-Sampling Network (PU-net) were proposed. The novel contribution of this article was as follows:

(1) In order to obtain the optimal solution of the initial parameters of the Kalman filter quickly and accurately, GOA took the sum of the longitude and latitude error variances as the fitness function to find the optimal solution to iterate the system noise variance matrix Q and the measurement noise variance matrix R. Based on the Q and R matrices from GOA, the positioning error of the UAV-borne LiDAR scanning system was greatly reduced; meanwhile, the computing burden was decreased, and the accuracy of point cloud data acquisition was improved;
(2) The density clustering algorithm was improved by the linear decreasing particle swarm optimization (PSO) algorithm. In order to solve the problem of sparse density, in this paper, a pre-trained PU-net was used for enhancing the point cloud density of the point cloud data obtained by UAV-borne LiDAR. Experiments have verified that the enhanced point cloud data retain the features of the raw point cloud data.

2. Framework of the Proposed Method

For UAV-borne LIDAR point cloud data enhancement, the integrated navigation algorithm based on GOA and the point cloud density enhancement algorithm based on neural network were shown in Figure 1. The specific implementation steps were as follows:

(1) Firstly, the GPS/INS integrated navigation information fusion framework was built by the Kalman filter. Then, the GOA was used to optimize the initial parameters of the Kalman filter, and the optimal solution was returned to the Kalman filter to estimate the output state of the GPS/INS integrated navigation system;
(2) After data acquisition, the point cloud data was transformed into the WGS-84 coordinate system from the LiDAR coordinate system. Then the point cloud data was filtered for removing the irrelevant points. Finally, the two-dimensional point cloud data was transformed into a three-dimensional point cloud according to the flight trajectory of UAV;
(3) The PSO algorithm was employed to optimize the density clustering algorithm for crop point cloud segment. Then the crop point cloud data was fed into the trained network to obtain the crop point cloud data after the point cloud density was enhanced. Finally, the point cloud data was reconstructed in three dimensions.

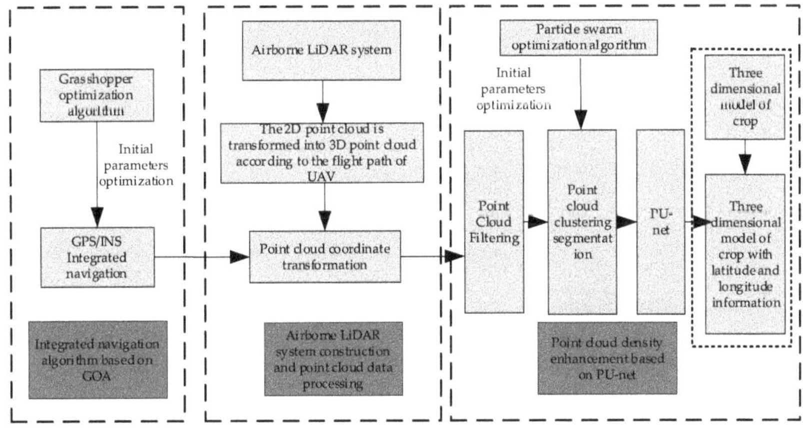

Figure 1. Framework of the proposed method. There are 3 steps: Integrated navigation algorithm based on grasshopper optimization algorithm (GOA), Unmanned Aerial Vehicle (UAV)-borne Light

Detection and Ranging (LiDAR) system construction and point cloud data processing, and point cloud density enhancement based on the Point Cloud Up-Sampling Network of the whole work of this paper.

3. Integrated Navigation Enhancement Based on GOA

In GPS/INS integrated navigation, the Kalman filter was usually used for fusing information of GPS and INS. The better the filter of integrated navigation was, the better it can integrate the advantages of each sensor. Since the GOA is not affected by the nonlinearity or scale, compared with other global optimization algorithms, it can find more effective optimal solutions with faster convergence speed [24].

3.1. Framework of GPS/INS Integrated Navigation

GPS and INS have complementary advantages. By combining GPS and INS in an appropriate way, their own drawbacks could be overcome, and the navigation accuracy would significantly improve. In order to improve the robustness of the system and reduce the complexity of the system, this paper adopted the loose combination mode to fuse the information of GPS and INS, as shown in Figure 2.

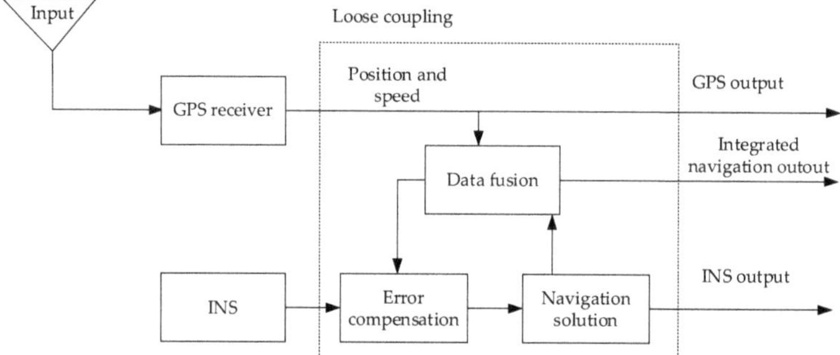

Figure 2. Framework of global positioning system (GPS)/ inertial navigation system (INS) integrated navigation.

The state variables of the integrated navigation filter were composed of the random drift error, velocity error, and position error of INS gyroscope, as well as errors caused by the attitude angle and accelerometer. The equation of the state was as follows:

$$\hat{X} = FX + Gw \tag{1}$$

The state variables of the INS integrated navigation system include longitude error, latitude error, velocity error in east and north directions, attitude angle error in east, north, and sky directions, random drift, and accelerometer bias error. The 13-dimensional equation of the state was as follows:

$$\hat{X}_I = F_I X_I + G_I \omega_I \tag{2}$$

The clock error of the receiver equivalent distance δt_u and equivalent range error of the receiver clock frequency error δt_{ru} were the GPS state errors; the equation of the state can be described as follows:

$$\hat{X}_G = F_G X_G + G_G w_G \tag{3}$$

The measurement equation of the position velocity combination was expressed as follows:

$$\begin{bmatrix} L_{INS} \\ \lambda_{INS} \\ h_{INS} \end{bmatrix} = \begin{bmatrix} L_t + \delta L \\ \lambda_t + \delta\lambda \\ h_t + \delta h \end{bmatrix} \quad (4)$$

where L_{INS}, λ_{INS}, and h_{INS} were the longitude, latitude, and altitude calculated by inertial navigation. L_t, λ_t, and h_t were the real longitude, latitude, and height of the carrier. δL, $\delta\lambda$, and δh were the longitude and latitude error and altitude error calculated by INS.

The position measurement information of the GPS receiver was expressed by subtracting the corresponding error from the real value:

$$\begin{bmatrix} L_{GPS} \\ \lambda_{GPS} \\ h_{GPS} \end{bmatrix} = \begin{bmatrix} L_t - \frac{N_n}{R} \\ \lambda_t - \frac{N_e}{R\cos L} \\ h_t - N_u \end{bmatrix} \quad (5)$$

where N_n, N_e, and N_u were the position error of the GNSS receiver along the east, north, and sky directions. Then the external measurement position error can be defined as:

$$Z_{pos} = \begin{bmatrix} \lambda_{INS} - \lambda_{GPS} \\ L_{INS} - L_{GPS} \\ h_{INS} - h_{GPS} \end{bmatrix} \quad (6)$$

The position measurement equation was as follows:

$$Z_{POS} = H_{POS}(t)X(t) + V_{POS}(t) \quad (7)$$

The velocity measurement information equation of INS can be expressed as follows:

$$\begin{bmatrix} v_{eINS} \\ v_{nINS} \\ v_{uINS} \end{bmatrix} = \begin{bmatrix} v_e + \delta v_e \\ v_n + \delta v_n \\ v_u + \delta v_u \end{bmatrix} \quad (8)$$

where v_{eINS}, v_{nINS}, and v_{uINS} were the east speed, north speed, and sky speed calculated by the inertial navigation system. v_e, v_n, and v_u were the velocity of the object in the east, north, and sky directions.

The velocity information provided by GPS can be expressed by the difference between the real value and the corresponding error in the geographical coordinate system:

$$\begin{bmatrix} v_{eGPS} \\ v_{nGPS} \\ v_{uGPS} \end{bmatrix} = \begin{bmatrix} v_e - M_e \\ v_n - M_n \\ v_u - M_u \end{bmatrix} \quad (9)$$

The speed error of the external measurement can be defined as:

$$Z_{vel} = \begin{bmatrix} v_{eINS} - v_{eGPS} \\ v_{nINS} - v_{nGPS} \\ v_{uINS} - v_{uGPS} \end{bmatrix} \quad (10)$$

The velocity measurement equation was as follows:

$$Z_{vel} = H_{vel}(t)X(t) + V_{vel}(t) \quad (11)$$

The integrated measurement equation of the GPS/INS integrated navigation system was obtained as follows:

$$Z(t) = \begin{bmatrix} \lambda_{INS} - \lambda_{GPS} \\ L_{INS} - L_{INS} \\ h_{INS} - h_{INS} \\ v_{eINS} - v_{eGPS} \\ v_{nINS} - v_{nGPS} \\ v_{uINS} - v_{uINS} \end{bmatrix} = H(t)X(t) + V(t) \tag{12}$$

3.2. Grasshopper Optimization Algorithm (GOA)

GOA is a swarm intelligence optimization algorithm designed by Australian scholar Shahrzad [18] in 2017, simulating the predatory migration behaviors of the grasshopper population in nature. The main characteristics of larval stage grasshoppers were slow movement and small stride. In contrast, long distances and sudden movements were essential features of adult grasshoppers. Searching for food sources was another important feature of the grasshopper swarm. The mathematical model of grasshopper swarm behavior was as follows:

$$X_i = \sum_{\substack{j=1 \\ j \neq i}}^{N} s\left(\left|x_j - x_i\right|\right)\frac{x_j - x_i}{d_{ij}} - g\hat{e}_g + u\hat{e}_w \tag{13}$$

where X_i represented the position of i_{th} grasshopper. N represents the amount of the grasshopper swarm. s was interaction force between grasshoppers. d_{ij} was the distance between the i_{th} grasshopper and j_{th} grasshopper. The function of the interaction force (attraction and repulsion shown in Figure 3.) *s* between grasshoppers was defined as follows:

$$s(d) = fe^{\frac{-d}{l}} - e^{-d} \tag{14}$$

where d was the distance between two grasshoppers. f represented the interaction force (attraction and repulsion), l represented the unit of interaction force length. The modified form of Equation (10) was as follows:

$$X_i^{dim} = c * r\left(\sum_{\substack{j=1 \\ j \neq i}}^{N} c\frac{ub_{dim} - lb_{dim}}{2}A\frac{x_j - x_i}{d_{ij}}\right) + \hat{T}_{dim} \tag{15}$$

$$A = s\left(\left|x_j^d - x_i^d\right|\right) \tag{16}$$

where ub_{dim} was the upper boundary of the dim_{th} dimension, lb_{dim} was the lower boundary, and $\hat{T}_{dim}$ was the target value of the dim_{th} dimension. c represented a self-adaption ratio which is defined as follows:

$$c = c_{max} - l\frac{c_{max} - c_{min}}{L} \tag{17}$$

where c_{max} was the max value, while c_{min} was the minimum value; l represented the current iteration, while L is the max iteration.

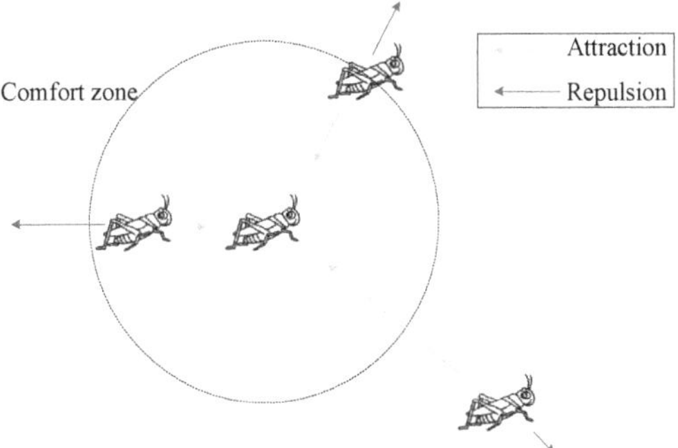

Figure 3. Interaction among grasshoppers.

3.3. Kalman Filter Optimization Using GOA

GOA was employed to optimize the initial parameters Q and R of the Kalman filter. Q was the variance matrix of the system noise sequence, which affected the filtering performance and parameter estimation accuracy of the Kalman filter algorithm; R was the variance matrix of the measurement noise sequence, which was related to the correction speed of filtering and the stability of the filtering process. Through the optimization of the initial parameters, the optimal position of the grasshopper was as close to the real value as possible, thus the accuracy of the Kalman filter algorithm was improved greatly. The sum of the variances of longitude and latitude errors was selected as the fitness function. The pipelines of the Kalman filter optimization was shown in Figure 4.

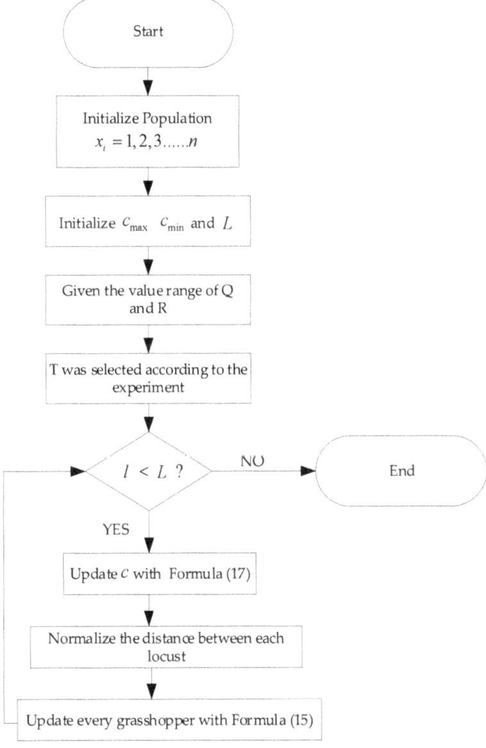

Figure 4. The pipelines of Kalman filter optimization.

4. UAV-Borne LiDAR System Construction and Point Cloud Data Processing

The UAV-borne LiDAR system was mainly composed of the LiDAR sensor, GPS/INS integrated navigation module, and UAV carrier. After the point cloud data were collected by the UAV-borne LiDAR scanning system, it was necessary to transform the point cloud data from the LiDAR coordinate system to the WGS-84 coordinate system.

4.1. UAV-Borne LiDAR System

In view of the application of LiDAR in the agricultural field, this paper selected RPLiDAR-a2 of SLAMTEC®, which was a single-line two-dimensional LiDAR, which greatly reduced the cost of equipment. The WitMotion® GPS/ Inertial Measurement Unit (IMU) was selected to collect the navigation and positioning information of UAV. In order to supply electricity to the LiDAR and the IMU, a small pad was used to combine the system. DJI® M100 UAV was selected as the carrier, which was a lightweight UAV. Finally, the UAV-borne LiDAR scanning system was completed, as shown in Figures 5 and 6.

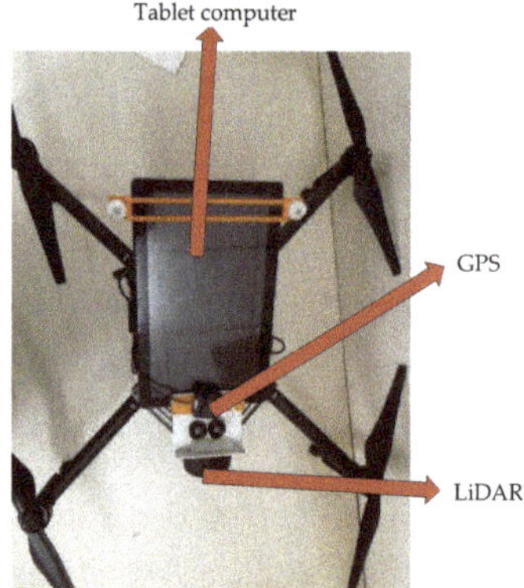

Figure 5. Top view of UAV-borne LiDAR scanning system.

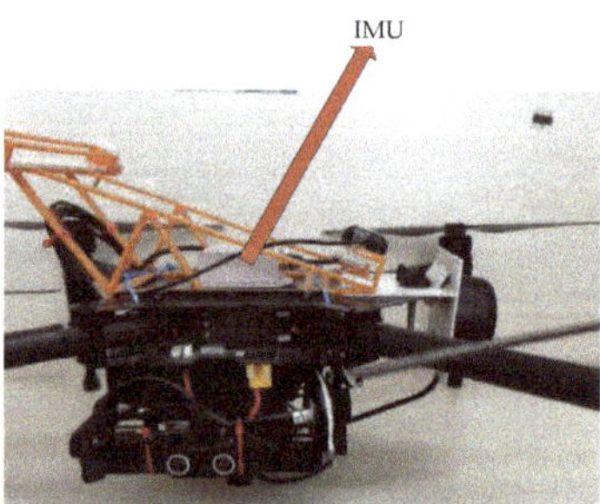

Figure 6. Side view of UAV-borne LiDAR system.

Due to the single scanning line of RPLiDAR-a2, the collected point cloud data was a two-dimensional point cloud, which only contained the data of Y and Z coordinates. In order to generate three-dimensional point cloud data, according to the light trajectory and speed of UAV, the data of the X-axis were inversely solved according to IMU.

4.2. Coordinate Transformation of Point Cloud Data

It was assumed that the coordinate of the laser foot point in the instantaneous laser coordinate system was $(x_{SL}, y_{SL}, z_{SL})^T$:

$$\begin{bmatrix} x_{SL} \\ y_{SL} \\ z_{SL} \end{bmatrix} = \begin{bmatrix} 0 \\ 0 \\ \rho \end{bmatrix} \tag{18}$$

There was an angle θ between the instantaneous laser coordinate system and the laser scanning reference coordinate system. The coordinates of the raw point cloud data in the laser scanning reference coordinate system were $(x_L, y_L, z_L)^T$:

$$\begin{bmatrix} x_L \\ y_L \\ z_L \end{bmatrix} = R_L \begin{bmatrix} x_{SL} \\ y_{SL} \\ z_{SL} \end{bmatrix} \tag{19}$$

where R_L represented the rotation matrix of scan angle θ.

Generally, the error between the different coordinate axes was analyzed and unified to be the placement error angle α, β, and γ. The deviation of the coordinate origin and reference center was treated as error $t_L = (\Delta x_I^L, \Delta y_I^L, \Delta z_I^L)^T$. The coordinates of the raw point cloud in the inertial platform reference coordinate system were $(x_I, y_I, z_I)^T$:

$$\begin{bmatrix} x_I \\ y_I \\ z_I \end{bmatrix} = R_M \begin{bmatrix} x_L \\ y_L \\ z_L \end{bmatrix} + \begin{bmatrix} \Delta x_I^L \\ \Delta y_I^L \\ \Delta z_I^L \end{bmatrix} \tag{20}$$

where $R_M = R(\gamma) \times R(\beta) \times R(\alpha)$.

In the integrated navigation system, there were also positioning errors between GPS and INS. During the installation process, there was a certain deviation between the center of GPS and the reference center of inertial components. The deviation was defined as $t_G = (\Delta x_I^G, \Delta y_I^G, \Delta z_I^G)^T$; it can be measured offline. INS measured the attitude information of the carrier in real time, including roll angle (R), pitch angle (P), and heading angle (H), and got the velocity and position information of the carrier through integration and feed back to the user in real time. At the same time, the coordinate rotation matrix was obtained by calculating the attitude information feedback from IMU. The coordinates of the raw point cloud data in the local reference coordinate system were $(x_{LH}, y_{LH}, z_{LH})^T$:

$$\begin{bmatrix} x_{LH} \\ y_{LH} \\ z_{LH} \end{bmatrix} = R_N \begin{bmatrix} x_I \\ y_I \\ z_I \end{bmatrix} - \begin{bmatrix} \Delta x_I^G \\ \Delta y_I^G \\ \Delta z_I^G \end{bmatrix} \tag{21}$$

where $R_N = R(H) \times R(P) \times R(R)$.

Finally, point cloud data should be transformed to WGS-84 coordinate system. The rotation matrix R_w was a function of the latitude and longitude of the region. The point cloud could be transformed into the WGS-84 coordinate system by rotating the matrix:

$$\begin{bmatrix} x_{84} \\ y_{84} \\ z_{84} \end{bmatrix} = R_W \begin{bmatrix} x_{LH} \\ y_{LH} \\ z_{LH} \end{bmatrix} + \begin{bmatrix} x_{84} \\ y_{84} \\ z_{84} \end{bmatrix}_{\text{Antenna phase center}} \tag{22}$$

Based on the above analysis, the coordinates of the raw point cloud in WGS-84 coordinate system were as follows:

$$\begin{bmatrix} x_{84} \\ y_{84} \\ z_{84} \end{bmatrix} = R_W R_N \left[R_M R_L \begin{bmatrix} 0 \\ 0 \\ \rho \end{bmatrix} + \begin{bmatrix} \Delta x_I^L \\ \Delta y_I^L \\ \Delta z_I^L \end{bmatrix} - \begin{bmatrix} \Delta x_I^G \\ \Delta y_I^G \\ \Delta z_I^G \end{bmatrix} \right] + \begin{bmatrix} x_{84} \\ y_{84} \\ z_{84} \end{bmatrix}_{\text{Antenna phase center}} \tag{23}$$

In addition, in order to get the data containing the latitude and longitude information, it was necessary to convert the three-dimensional rectangular coordinates (XYZ) into geodetic coordinates (λLH). The conversion formula was as follows:

$$L = \arctan(Y/X) \tag{24}$$

$$\lambda = \arctan\left[Z/\sqrt{X^2 + Y^2} * \left(1 - e^2 N(N + H)\right)^{-1}\right] \tag{25}$$

$$H = \sqrt{X^2 + Y^2}/\cos\lambda - N \tag{26}$$

where X, Y, and Z were three-dimensional rectangular coordinate components, L, λ, and *H* were earth longitude, latitude, and height. e represented the first eccentricity of the ellipsoid. $N = a/\sqrt{1 - e^2 \sin^2 \lambda}$ was the radius of curvature in the prime vertical, and the a was the semimajor axis of the ellipsoid.

5. Point Cloud Density Enhancement Based on PU-Net

Point cloud density enhancement aims to increase the number of point clouds without destroying the characteristics of the original point cloud. By enhancing the density of point cloud, the richness and accuracy of 3D reconstruction of point cloud data can be improved without increasing equipment cost.

5.1. Point Cloud Preprocessing

After the coordinate transformation, it was necessary to remove irrelevant noise points from the original point cloud data. In this paper, the statistical filtering method was selected for filtering. Statistical filtering calculated the distance between each point and its surrounding neighborhood points: According to neighborhood points and the standard deviation threshold, the points which did not meet the requirements were identified as outlier points.

5.2. Point Cloud Density Clustering Segmentation Method Based on PSO Algorithm

After filtering in the previous section, the point cloud data obtained included ground and crop points. In order to separate the ground and crop, the density clustering algorithm was used for separating, and the linear decreasing PSO algorithm was employed to improve the density clustering algorithm.

DBSCAN was a density-based clustering algorithm proposed by ester et al. It was designed to achieve clustering and segmentation of arbitrary data in spatial and non-spatial high-dimensional databases. By calculating the distance between each point and its neighborhood, the points with similar density were classified into the same class.

ε-Neighborhood of arbitrary point p is defined as follows:

$$N_{eps} = \{q \in D/\text{dist}(p, q) < \text{Eps}\} \tag{27}$$

where D represented the target's database, point p was the core point if the ε-neighborhood contained at least the minimum number of points. The core point was defined as follows:

$$N_{eps}(P) > \text{MinPts} \tag{28}$$

where Eps and MinPts were the neighborhood radius and minimum number of points in the ε-neighborhood of the core point. Through the calculation of all the points, the density of the points meeting the set conditions was similar, and they were classified as the same kind of points. All points were calculated until the termination condition was met. Because the selection of these two initial parameters mainly depended on human experience and continuous experiments, in order to improve the efficiency, this paper employed the linear decreasing PSO algorithm to iteratively optimize the two initial parameters to determine the optimal solution of the initial parameters.

The PSO algorithm was a swarm intelligent bionic optimization algorithm: Every point in PSO was stamped by a position P_i, the optimal value of swarm was P_g, every point's velocity was V_i, and each particle's velocity and position update formula was shown as follows:

$$V_i^{t+1} = \omega V_i^t + c_1 r_1 (P_i - X_i^t) + c_2 r_2 (P_g - X_i^t) \tag{29}$$

$$X_i^{t+1} = X_i^t + V_i^{t+1} \tag{30}$$

where t represented the iterations, V_i^{t+1} was the velocity of the $t+1$ iteration, ω represented the inertia weight. c_1, $c_2 \in [0,1]$ were the generating functions of a random number, and X_i^{t+1} was the $t+1$ iteration position.

In the initial stage of the algorithm, the inertia weight ω was given a larger value to enhance the global exploration ability of the algorithm. With the continuous exploration process, the value of ω could be linearly reduced to enhance the local search and optimization ability of the algorithm, so that particles could gradually approach the optimal solution in the neighborhood of the optimal value. The value of ω was expressed as follows:

$$\omega = \omega_{min} + \frac{(\omega_{max} - \omega_{min}) \times (max_{iter} - t)}{max_{iter}} \tag{31}$$

where ω_{max} represented the max inertia coefficient, while ω_{min} represented the minimum. max_{iter} was the max iteration.

PSO optimization pipelines are shown in Figure 7:

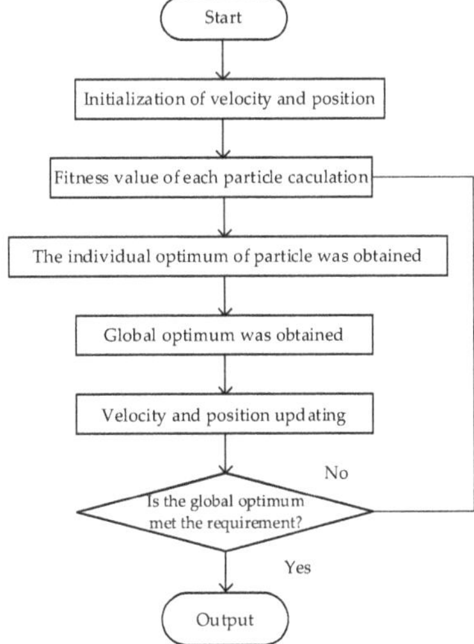

Figure 7. Flow chart of particle swarm optimization (PSO) algorithm.

5.3. Point Cloud Density

The point cloud data acquired by UAV-borne LiDAR was 3-dimensional, which was usually expressed by plane density. The point cloud density in national standards was defined as the normal direction of elevation density [22]:

$$\rho = \frac{N - \sum_{i=0}^{k} N_i}{A - \sum_{i=0}^{k} A_i} \tag{32}$$

where ρ represented the point cloud density, and N represented the total amount of points. Furthermore, N_i was the i_{th} field number of points, and k was the number of scanning area. A was the measure of the scanning area.

The point cloud density calculated by the above formula was the average point cloud density in the scanning area. In order to further quantify the point density distribution in the scanning area, the reciprocal of the area affected by a single point was taken as the density of the point:

$$d' = 1/A' \tag{33}$$

The main factors affecting the density of the point cloud were the LiDAR design parameters, system characteristics, carrier, and scanning area.

The distance between points directly reflected the density of point clouds. For the pulse laser ranging sensor, the distance from the heading point to the edge point determined the density distribution of point cloud. The design of aerial photography parameters was referred from the distance between heading point and edge point. The parameters of the sensor, such as scanning frequency, scanning bandwidth, laser emission frequency, field of view angle, flight height and flight speed, and the distance between point clouds, were considered to design the LiDAR system.

5.4. Density Enhancement Using Point Cloud Up-Sampling Network (PU-Net)

Point cloud up-sampling was essentially similar to the problem of image super-resolution [25]. Yu et al. [26] proposed a data-driven point cloud sampling neural network (PU-net), which was applied to patch level. At the same time, a joint loss function was used for guiding the up-sampled points to retain a uniformly distributed object surface. The main idea of PU-net was capturing the features of each point in every level, then multi-convolution blocks were extended in feature hidden layers, and finally, the hidden layers were divided into multi-level features and up-sampled into the point cloud. The pre-trained PU-net was employed for enhancing the density of the crop point cloud in this paper. There were 4 steps of PU-net for up-sampling of point cloud: Slice extraction, point feature embedding, feature extending, and point coordinate reconstruction. It is noteworthy that a series of sub-pixel convolution layers were employed to extend the features of neighborhood points in the feature extending stage:

$$f' = \mathrm{Reshape}([\mathrm{conv}_1^2(\mathrm{conv}_1^1(f)), \ldots, \mathrm{conv}_r^2(\mathrm{conv}_r^1(f))]) \tag{34}$$

where $\mathrm{conv}_i^1(*)$ and $\mathrm{conv}_i^2(*)$ were the convolution layer with 1*1 kernel size for adjusting the channels of patch feature f. The output, combined feature f′, was born with properties of spatial characteristic which could retain the information of raw data. After coordinate reconstruction, the density elevation in last section was employed to show the enhancement effect directly.

6. Results

In order to verify the stability of the UAV-borne LiDAR scanning system, the experiments were carried out in integrated navigation testing, point cloud density clustering, and point cloud density enhancement.

6.1. GOA-Based Integrated Navigation

The integrated navigation experiment was carried out in MATLAB2017 (a). It was supposed that the initial longitude of UAV $L = 40^{\circ}$, the initial latitude $\lambda = 116^{\circ}$, and the velocity of UAV was 1m/s. The filtering period was $T = 1s$. The initial value of filtering covariance was as follows:

$$\begin{aligned} p(0) = \text{diag}[&(0.00254^{\circ})^2, (0.00446^{\circ})^2, (0.1\text{m/s})^2, \\ &(0.1\text{m/s})^2, (1^{\circ})^2, (1^{\circ})^2, (1^{\circ})^2, (1^{\circ})^2, \\ &(1^{\circ})^2, (1^{\circ})^2, (10^{-5})g, (10^{-5})g, (10^{-5})g] \end{aligned} \tag{35}$$

The initial value of integrated navigation was:

$$\begin{aligned} X(0) = [&0\text{m}, 0\text{m}, 0.5\text{m/s}, 0.5\text{m/s}, 0.08^{\circ}, 0.08^{\circ}, \\ &0.08^{\circ}, 0.16^{\circ}, 0.1^{\circ}/\text{h}, 0.1^{\circ}/\text{h}, 0.1^{\circ}/\text{h}, \\ &0.1^{\circ}/\text{h}, (10^{-5})g, (10^{-5})g, (10^{-5})g] \end{aligned} \tag{36}$$

The initial parameters Q and R obtained when the sum of longitude and latitude errors were used as the fitness function feeding into the system. The GOA based integrated navigation error diagram was shown in Figures 8 and 9, PSO based integrated navigation error diagram was shown in Figures 10 and 11.

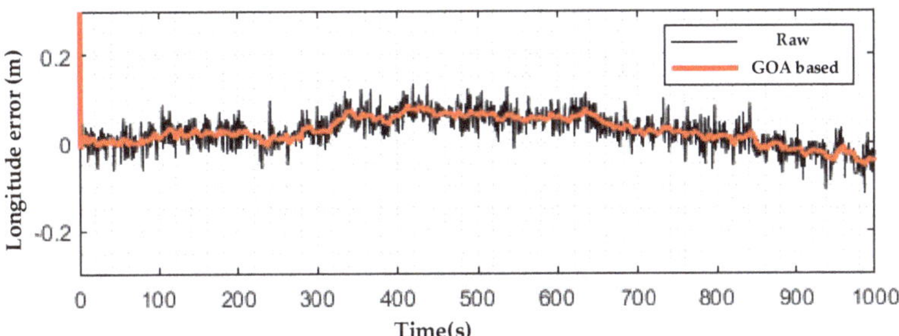

Figure 8. Longitude error of GOA-based integrated navigation.

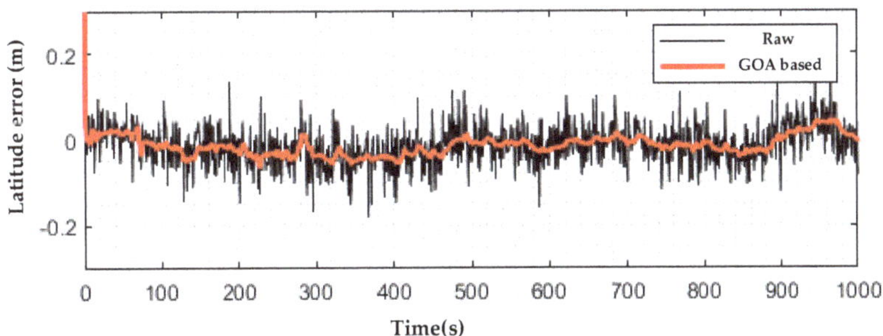

Figure 9. Latitude error of GOA-based integrated navigation.

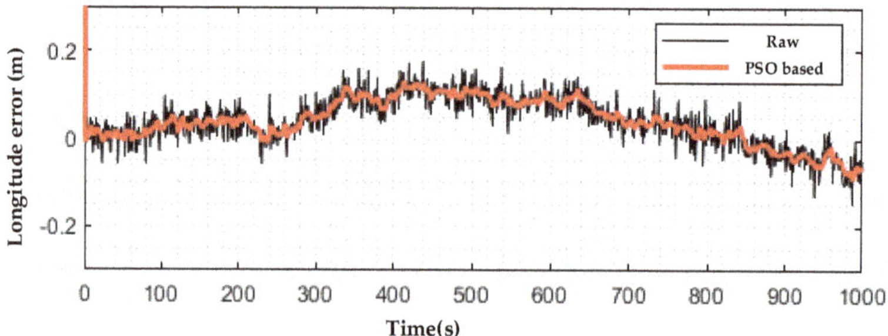

Figure 10. Longitude error of PSO-based integrated navigation.

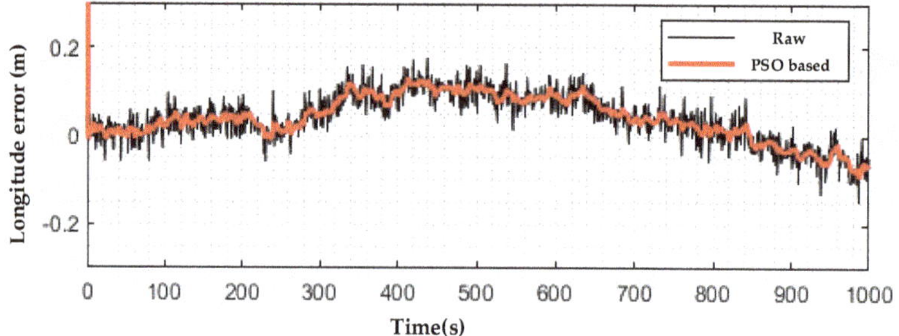

Figure 11. Latitude error of PSO-based integrated navigation.

As shown in Table 1, both the longitude L variance error and latitude λ variance error were reduced greatly compared with raw integrated navigation. It is worth noting that the GOA-based integrated navigation showed better performance than the PSO-based integrated navigation; it is suspected that the ability to find the optimal of GOA was better than for PSO in the integrated navigation task. For more precise optimal Q and R, the GOA-based integrated navigation was employed for the UAV-borne LiDAR scanning system.

Table 1. Error of integrated navigation with different enhancement methods.

	Longitude L Variance Error	Latitude λ Variance Error
GOA-based integrated navigation	0.00046	0.00034
PSO-based integrated navigation	0.00087	0.00079
Raw integrated navigation	0.0091	0.0047

6.2. PSO-Based Point Cloud Clustering Segmentation

The meta-heuristic algorithm was employed to optimizing the clustering segmentation key elements: Eps and MinPts. The Davies–Bouldin index (DBI) was employed as the fitness function for the joint optimization of Eps and MinPts:

$$\mathrm{DBI} = \frac{1}{k}\sum_{i=1}^{k} \max_{j \neq i}\left(\frac{\mathrm{avg}(C_i) + \mathrm{avg}(C_j)}{d_{cen}(\mu_i, \mu_j)}\right) \tag{37}$$

where avg(∗) represented the average distance between samples in cluster, $d_{cen}(\mu_i, \mu_j)$ was the center distance of cluster μ_i, and cluster μ_j. avg(*) was defined as follows:

$$avg(C) = \frac{2}{|C|(|C|-1)} \sum_{1 \le i \le j \le |C|} dist(x_i, x_j) \tag{38}$$

where $dist(x_i, x_j)$ was the distance of sample x_i and sample x_j.

Crop point cloud data collection environment was shown in Figure 12.

Figure 12. Crop point cloud raw data collection environment.

A series of crop point cloud raw data was collected from the UAV-borne LiDAR scanning system; the meta-heuristic-based DBSCAN algorithms were initialized by crop raw data and iterated 200 times to find the optimal Eps and MinPts. The fitness iteration curve is shown in Figure 13.

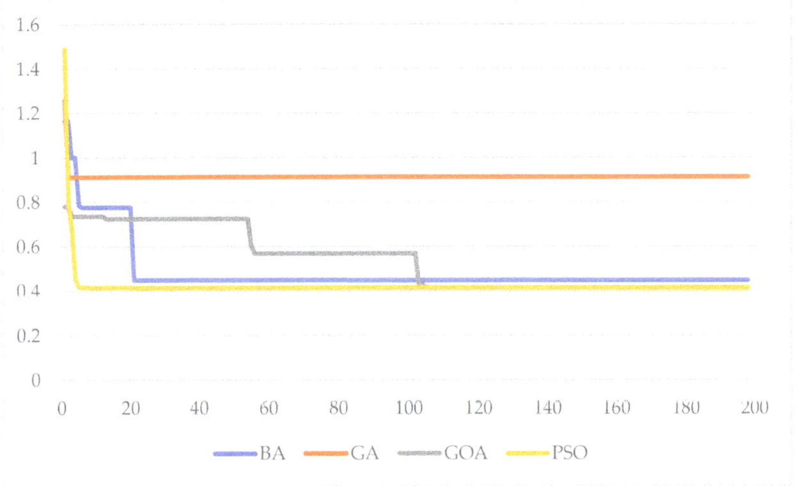

Figure 13. Iteration curve of fitness of the meta-heuristic-based density-based spatial clustering of applications with noise (DBSCAN) algorithms.

For fair comparing with different meta-heuristic algorithms, the BA-based DBSCAN algorithm [17], GA-based DBSCAN algorithm [14], GOA-based DBSCAN algorithm [18], and PSO-based DBSCAN algorithm [15] were initialized with 20 population-sizes, 2 dimensions, and 200 iterations. In particular, the GA-based DBSCAN algorithm was initialized with a 0.7 crossover rate, 0.5 select rate, and 0.01 variation rate. During several times of testing, Figure 13 shows that the GOA and PSO found the best fitness 0.416032. It is suspected that global optimization was 0.416032 under further testing with different fine tuning. Under the same fitness 0.416032, the Eps = 0.8147 and MinPts = 13. In the clustering segmentation, the BA-based DBSCAN algorithm, GA-based DBSCAN algorithm, GOA-based DBSCAN algorithm, PSO-based DBSCAN algorithm, and the K-means clustering were employed for comparison; the K-means was the default tuning. The 200 iterations of time consumption of the BA-based DBSCAN algorithm, GA-based DBSCAN algorithm, GOA-based DBSCAN algorithm, and PSO-based DBSCAN algorithm were 9669 s, 910 s, 1610 s, and 749 s, respectively. The comparison clustering results are shown in Figure 14a–e. Shown in Figure 14f, the crop point cloud is reconstructed in 3-dimensions.

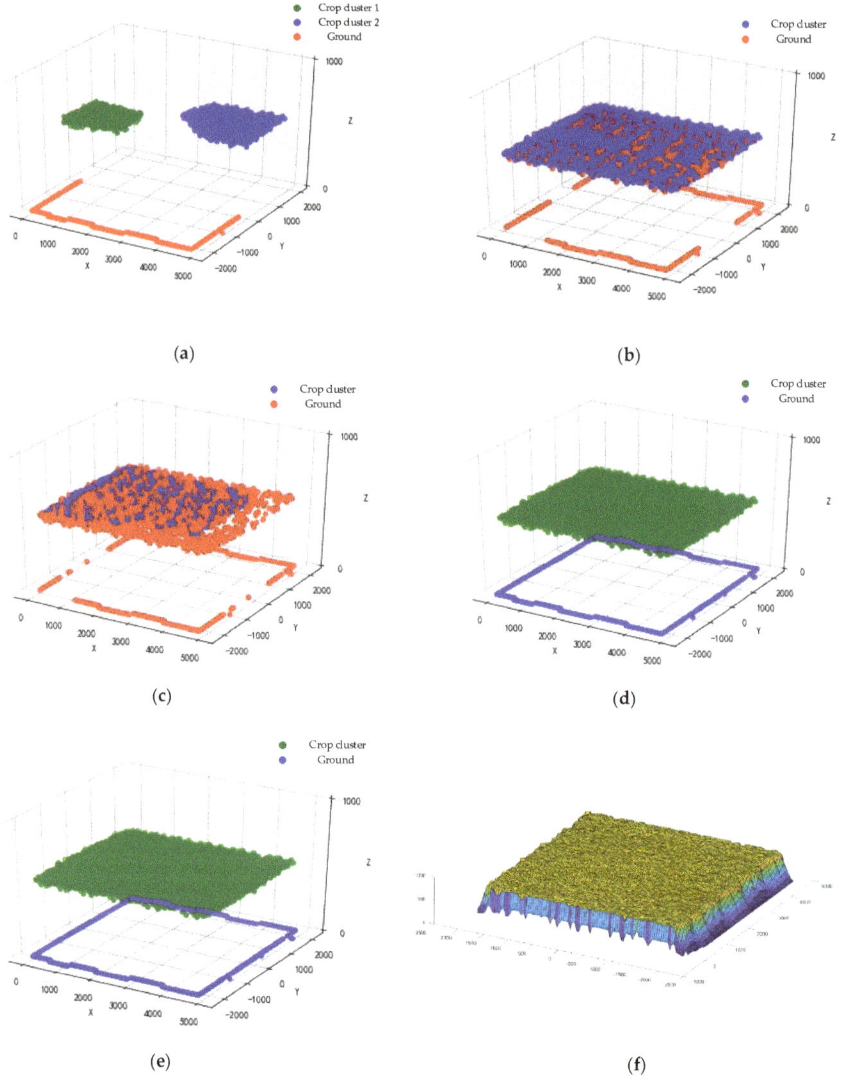

Figure 14. Clustering segmentation result of crop and ground. (**a**) K-means clustering; (**b**) BA-based DBSCAN (Eps = 0.2145, MinPts = 18); (**c**) GA-based DBSCAN (Eps = 0.1006, MinPts = 8); (**d**) PSO-based DBSCAN (Eps = 0.8147, MinPts = 13); (**e**) GOA-based DBSCAN (Eps = 0.8147, MinPts = 13); (**f**) 3D reconstruction of crop field point cloud.

As shown in Figure 14a–c, 3 kinds of clustering results were given from the default tuning of K-means, and the ground cluster could not be separated from the crop cluster using the BA-based DBSCAN algorithm or GA-based DBSCAN algorithm. Under the same fitness 0.416032, the Eps = 0.8147 and MinPts = 13 were obtained using the GOA-based DBSCAN algorithm and PSO-based DBSCAN algorithm. Considering that PSO converges faster and takes less time in iterations, the PSO-based DBSCAN algorithm was employed for crop point cloud clustering segmentation.

6.3. Density Enhancement with PU-Net

After the clustering segmentation of crop point cloud, the crop point cloud was enhanced by a pre-trained PU-net. The PU-net was trained with 60 sets of variation of point clouds from "Visionair". Datasets were divided into train sets and validation sets by 60% and 40%. In the fine-tune training period, the Monte Carlo stochastic sampling method was employed for sampling 5000 points from every object as the input. Each input represented the global information of each object; fine-tune training aimed at enhancing the global information collection performance of the pretrained model. The visualization of raw crop point cloud data was shown in Figure 15. The sparse part of the point cloud was mainly concentrated on the X-axis, which was the direction of the UAV-borne LiDAR scanning system to generate the 3D point cloud by moving forward.

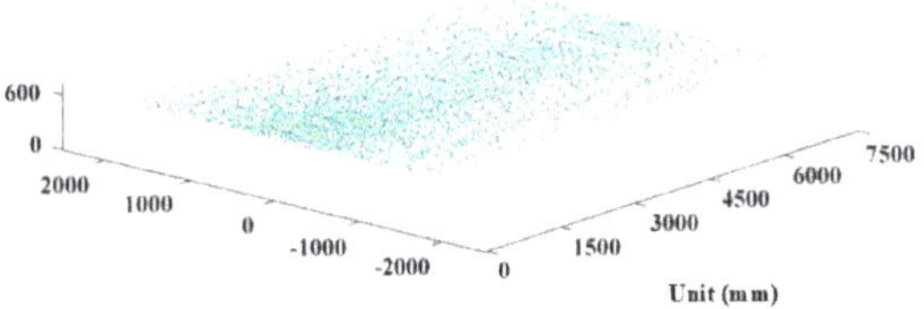

Figure 15. Visualization of raw crop point cloud data.

The raw crop point cloud was fed into the fine-tune trained PU-net. The 3D reconstruction of the crop point cloud with and without density enhancement is shown in Figure 16.

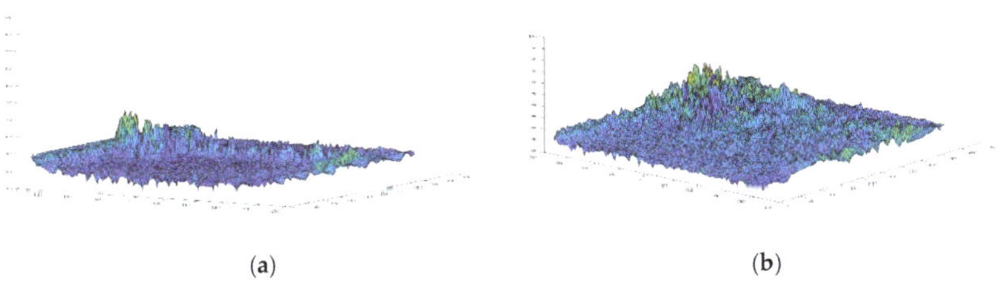

(a) (b)

Figure 16. 3D reconstruction of crop point cloud with and without density enhancement. (**a**) The raw crop point cloud; (**b**) The crop point cloud with density enhancement.

The total number of points of the raw crop point cloud were 10,000, the density of the raw crop point cloud was 1250 points per square meter. After density enhancement, the total number of points of the crop point cloud were 40,000, the density of the crop point cloud was 5000 points per square meter, and the number and density of the point cloud had been significantly increased. Compared with the raw crop point cloud, the enhanced crop point cloud contained more information, and also retained the shape and contour characteristics of the raw data. In order to further verify the reliability and effectiveness of the method, 50 sets of density enhancement tests were carried out and 5 sets of results were sampled randomly for statistical tests. The results were shown in Table 2.

Table 2. The statistical parameters of the enhanced point cloud and the raw data.

	Mean Value	Variance	Standard Deviation
Raw crop point cloud	558.64	171.60	13.10
1st set of the enhanced crop point cloud	568.84	298.48	17.27
2nd set of the enhanced crop point cloud	578.68	289.34	17.01
3rd set of the enhanced crop point cloud	554.32	278.56	16.69
4th set of the enhanced crop point cloud	565.44	325.08	18.03
5th set of the enhanced crop point cloud	563.56	317.51	17.82

7. Discussion

For integrated navigation testing, because the GOA and PSO showed better performance in fitness convergence velocity in the local field refer to Ref. [19]. Although this was an unfavorable feature for global optimization, for time varying and unpredictable problems, such as the integrated navigation system in this paper, better performance in the fitness convergence velocity in the local field tended to find the suboptimal solution quickly, which would reduce the error sharply. Shown as the GOA based integrated navigation error diagram (Figures 8 and 9) and the PSO based integrated navigation error diagram (Figures 10 and 11), both the longitude L variance error and latitude λ variance error were reduced greatly compared with raw integrated navigation. It is worth noting that the GOA-based integrated navigation showed better performance than the PSO-based integrated navigation; it is suspected that the ability to find the optimal of GOA was better than for PSO in the integrated navigation task. For more precise optimal Q and R, the GOA-based integrated navigation was employed for the UAV-borne LiDAR scanning system.

For point cloud density clustering segmentation, default-turning K-means clustering divided crop into two clusters and separates ground points clearly, but parts of data points were ignored for clustering. Different meta-heuristic-based DBSCAN algorithms shown different optimization capabilities according to Figure 13. GA algorithm was the weakest in point cloud density clustering, different variation rates and 800 iterations more were set in further fine-turning, the GA-based DBSCAN algorithms still could not find the lower fitness value which was close to 0.416032. The BA-based DBSCAN algorithm shown the better optimization capability but lower computing efficiency, 9669 s of time consumption make it worse for practical application. Both GOA-based DBSCAN algorithm and PSO-based DBSCAN algorithm found the same fitness 0.416032. It is suspected that global optimization was 0.416032 under further testing with different fine tuning. Considering that PSO converges faster and takes less time in iterations, the PSO-based DBSCAN algorithm was employed for crop point cloud clustering segmentation.

For point cloud density enhancement, the pre-trained PU-net [26] was trained by "Visionair" datasets, there were many objects point cloud dataset for learning features by Convolutional Neural Network (CNN). After point cloud density enhancement, the density of point cloud was greatly enhanced, and Table 2 shown that there was no gap in mean value between the raw crop point cloud and 5 sets of the enhanced crop point cloud, which indicated that the enhanced point cloud kept the height distribution of the raw data. A faint change of variance and standard deviation indicated that the 5 sets of the enhanced point cloud also kept the shape and contour of the raw data.

8. Conclusions

This paper developed a UAV-borne LiDAR scanning system with enhanced integrated navigation and dense point cloud optimization. The GOA was employed for optimizing the integrated navigation; the longitude and latitude variances of optimized integrated navigation were 0.00046 and 0.00034, respectively, and the longitude and latitude variances of raw integrated navigation were 0.0091 and 0.0047, respectively. The positioning accuracy of the UAV carrier was improved greatly. The PSO was employed to find the optimal key elements of DBSCN clustering segmentation, and accurate point cloud segmentation was obtained. At last, the fine-tuned PU-net was used for point cloud density enhancement. Compared with the raw point cloud, the density of the enhanced crop point cloud was 4 times more, and the statistical parameters showed that the enhanced crop point cloud retained the shape and contour characteristics of the raw data. In this paper, each meta-heuristic optimization algorithm had its own merits for the different problems, but we still have not found an optimization algorithm that can meet all the processing requirements. In the future, owing to the flexibility of a multi-rotor UAV in remote sensing operations in specific areas, a multi-rotor UAV equipped with a small LiDAR or optical image remote sensing system will be more suitable for remote sensing operations in the agricultural field [27,28].

Author Contributions: Conceptualization, J.C. and Z.Z.; methodology, J.C. and Z.Z.; software, Z.Z. and K.Z.; validation, Z.Z., K.Z., and S.W.; formal analysis, S.W. and Y.H.; investigation, S.W. and Y.H.; data curation, K.Z. and S.W.; writing—original draft preparation, J.C. and Z.Z.; writing—review and editing, K.Z., S.W., and Y.H.; supervision, J.C. and Y.H.; project administration, J.C. and Y.H.; funding acquisition, J.C. and Y.H. All authors have read and agreed to the published version of the manuscript.

Funding: This research was funded by the National Natural Science Foundation of China (Grant No. 51979275), by the National Key R&D Program of China (Grant Nos. 2017YFD0701003 from 2017YFD0701000, 2018YFD0700603 from 2018YFD0700600, and 2016YFD0200702 from 2016YFD0200700), by the Jilin Province Key R&D Plan Project (Grant Nos. 20180201036SF), by the Open Fund of Synergistic Innovation Center of Jiangsu Modern Agricultural Equipment and Technology, Jiangsu University (Grant No. 4091600015), by the Open Research Fund of State Key Laboratory of Information Engineering in Surveying, Mapping and Remote Sensing, Wuhan University (Grant No. 19R06), by the Open Research Project of the State Key Laboratory of Industrial Control Technology, Zhejiang University (Grant No. ICT20021), and by the Chinese Universities Scientific Fund (Grant Nos. 2020TC033 and 10710301).

Conflicts of Interest: The authors declare no conflict of interest.

References

1. McManamon, P.F. Review of LiDAR: A historic, yet emerging, sensor technology with rich phenomenology. *Opt. Eng.* **2012**, *51*, 060901. [CrossRef]
2. Vázquez-Arellano, M.; Griepentrog, H.W.; Reiser, D.; Paraforos, D.S. 3-D Imaging Systems for Agricultural Applications—A Review. *Sensors* **2016**, *16*, 618. [CrossRef] [PubMed]
3. Lovell, J.L.; Jupp, D.L.B.; Newnham, G.J. Measuring tree stem diameters using intensity profiles from ground-based scanning LiDAR from a fixed viewpoint. *ISPRS J. Photogramm. Remote Sens.* **2011**, *66*, 46–55. [CrossRef]
4. Strahler, A.H.; Jupp, D.L.B.; Woodcock, C.E. Retrieval of forest structural parameters using a ground-based LiDAR instrument (Echidna). *Can. J. Remote Sens.* **2008**, *34*, 426–440. [CrossRef]
5. Zheng, G.; Moskal, L.M.; Kim, S.H. Retrieval of effective leaf area index in heterogeneous forests with terrestrial laser scanning. *IEEE Trans. Geosci. Remote Sens.* **2013**, *51*, 777–786. [CrossRef]
6. Lai, X.D.; Liu, Y.S.; Li, Y.X. Application status and development of point cloud density characteristics of airborne LiDAR. *Geospat. Inf.* **2018**, *16*, 1–5.
7. Quan, W.; Gong, X.; Fang, J.; Li, J. Prospects of INS/CNS/GNSS Integrated Navigation Technology. In *INS/CNS/GNSS Integrated Navigation Technology*; Springer: Berlin/Heidelberg, Germany, 2015.
8. Fu, B.; Liu, J.; Wang, Q. Multi-sensor integrated navigation system for ships based on adaptive Kalman filter. In Proceedings of the 2019 IEEE International Conference on Mechatronics and Automation (ICMA), Tianjin, China, 4–9 August 2019.
9. Gao, Y.; Liu, S.; Atia, M.M.; Noureldin, A. INS/GPS/LiDAR integrated navigation system for urban and indoor environments using hybrid scan matching algorithm. *Sensors* **2015**, *15*, 23286–23302. [CrossRef] [PubMed]
10. Shabani, M.; Gholami, A.; Davari, N. Asynchronous direct Kalman filtering approach for underwater integrated navigation system. *Nonlinear Dyn.* **2015**, *80*, 71–85. [CrossRef]
11. Zhang, G.; Hsu, L.T. Intelligent GNSS/INS integrated navigation system for a commercial UAV flight control system. *Aerosp. Sci. Technol.* **2018**, *80*, 368–380. [CrossRef]
12. Yu, H.; Han, H.; Wang, J.; Xiao, H.; Wang, C. Single-frequency GPS/BDS RTK and INS ambiguity resolution and positioning performance enhanced with positional polynomial fitting constraint. *Remote Sens.* **2020**, *12*, 2374. [CrossRef]

13. Norouz, M.; Ebrahimi, M.; Arbabmir, M. Modified Unscented Kalman Filter for improving the integrated navigation system. In Proceedings of the 25th Iranian Conference on Electrical Engineering, Tehran, Iran, 2–4 May 2017.
14. Goldberg, D. *Genetic Algorithms in Search, Optimization and Machine Learing*; Addison-Wesley: Boston, MA, USA, 1989.
15. Eberhart, R.; Kennedy, J. A new optimizer using particle swarm theory. In Proceedings of the 6th International Symposium on Micro Machine and Human Science, Nagoya, Japan, 4–6 October 1995; pp. 39–43.
16. Colorni, A.; Dorigo, M.; Maniezzo, V. Distributed optimization by ant colonies. In Proceedings of the 1st European Conference on Artificial Life, Paris, France, 11–13 December 1991; pp. 134–142.
17. Yang, X. A new metaheuristic bat-inspired algorithm. In *Proceedings of Nature Inspired Cooperative Strategies for Optimization (NICSO 2010)*; Springer: Berlin/Heidelberg, Germany, 2010; pp. 65–74.
18. Saremi, S.; Mirjalili, S.; Lewis, A. Grasshopper optimisation algorithm: Theory and application. *Adv. Eng. Softw.* **2017**, *105*, 30–47. [CrossRef]
19. Feng, H.; Ni, H.; Zhao, R.; Zhu, X. An enhanced grasshopper optimization algorithm to the Bin packing problem. *J. Control Sci. Eng.* **2020**, *2020*, 3894987. [CrossRef]
20. Chen, Y.; Zhou, L.; Bouguila, N.; Wang, C.; Chen, Y.; Du, J. BLOCK-DBSCAN: Fast clustering for large scale data. *Pattern Recognit.* **2020**, *109*, 107624. [CrossRef]
21. Kedzierski, M.; Fryskowska, A. Methods of laser scanning point clouds integration in precise 3D building modelling. *Measurement* **2015**, *74*, 221–232. [CrossRef]
22. Rupink, B.; Mongus, D.; Zalik, B. Point density evaluation of airborne LiDAR datasets. *J. Univers. Comput. Sci.* **2015**, *21*, 587–603.
23. Huang, H.; Shi, W.U.; Gong, M. Edge-aware point set resampling. *ACM Trans. Graph.* **2013**, *32*, 1–12. [CrossRef]
24. Mafarja, I.A. Evolutionary population dynamics and grasshopper optimization approach for feature selection problems. *Knowl. Based Syst.* **2018**, *145*, 25–45. [CrossRef]
25. Ledig, C.; Theis, L.; Huszar, F. Photo-realistic single image super-resolution using generative adversarial network. In Proceedings of the IEEE Conference on Computer Vision and Pattern Recognition, Honolulu, HI, USA, 21–26 July 2017.
26. Yu, L.; Li, X.; Fu, C.W. PU-net: Point cloud upsampling network. In Proceedings of the IEEE Conference on Computer Vision and Pattern Recognition, Salt Lake City, UT, USA, 19–21 June 2018.
27. Zhang, Z.; Han, Y.; Chen, J.; Wang, S.; Wang, G.; Du, N. Information extraction of ecological canal system based on remote sensing data of unmanned aerial vehicle. *J. Drain. Irrig. Mach. Eng.* **2018**, *36*, 1006–1011.
28. Wang, S.; Han, Y.; Chen, J.; Pan, Y.; Cao, Y.; Meng, H. Weed classification of remote sensing ecological irrigation area by UAV based on deep learning. *J. Drain. Irrig. Mach. Eng.* **2018**, *11*, 1137–1141.

Article

Wavelength Selection Method Based on Partial Least Square from Hyperspectral Unmanned Aerial Vehicle Orthomosaic of Irrigated Olive Orchards

Antonio Santos-Rufo [1,*], Francisco-Javier Mesas-Carrascosa [2], Alfonso García-Ferrer [2] and Jose Emilio Meroño-Larriva [2]

[1] Department of Agronomy, University of Cordoba, Campus de Rabanales, 14071 Córdoba, Spain
[2] Department of Graphic Engineering and Geomatics, University of Cordoba, Campus de Rabanales, 14071 Córdoba, Spain; fjmesas@uco.es (F.-J.M.-C.); agferrer@uco.es (A.G.-F.); ir1melaj@uco.es (J.E.M.-L.)
* Correspondence: g02sarua@uco.es

Received: 13 September 2020; Accepted: 16 October 2020; Published: 19 October 2020

Abstract: Identifying and mapping irrigated areas is essential for a variety of applications such as agricultural planning and water resource management. Irrigated plots are mainly identified using supervised classification of multispectral images from satellite or manned aerial platforms. Recently, hyperspectral sensors on-board Unmanned Aerial Vehicles (UAV) have proven to be useful analytical tools in agriculture due to their high spectral resolution. However, few efforts have been made to identify which wavelengths could be applied to provide relevant information in specific scenarios. In this study, hyperspectral reflectance data from UAV were used to compare the performance of several wavelength selection methods based on Partial Least Square (PLS) regression with the purpose of discriminating two systems of irrigation commonly used in olive orchards. The tested PLS methods include filter methods (Loading Weights, Regression Coefficient and Variable Importance in Projection); Wrapper methods (Genetic Algorithm-PLS, Uninformative Variable Elimination-PLS, Backward Variable Elimination-PLS, Sub-window Permutation Analysis-PLS, Iterative Predictive Weighting-PLS, Regularized Elimination Procedure-PLS, Backward Interval-PLS, Forward Interval-PLS and Competitive Adaptive Reweighted Sampling-PLS); and an Embedded method (Sparse-PLS). In addition, two non-PLS based methods, Lasso and Boruta, were also used. Linear Discriminant Analysis and nonlinear K-Nearest Neighbors techniques were established for identification and assessment. The results indicate that wavelength selection methods, commonly used in other disciplines, provide utility in remote sensing for agronomical purposes, the identification of irrigation techniques being one such example. In addition to the aforementioned, these PLS and non-PLS based methods can play an important role in multivariate analysis, which can be used for subsequent model analysis. Of all the methods evaluated, Genetic Algorithm-PLS and Boruta eliminated nearly 90% of the original spectral wavelengths acquired from a hyperspectral sensor onboard a UAV while increasing the identification accuracy of the classification.

Keywords: olive tree; UAV; hyperspectral; classification; irrigation technique; PLS; wavelength selection

1. Introduction

The intensification of agricultural practices, including better seeds, extensive fertilizer use and irrigation techniques, has altered the dynamics between humans and environmental systems across the world [1]. Although these agricultural practices have allowed for increased food production, they have also caused significant environmental impact on many regions. Consequently, accurate and precise information is in high demand from Earth System Science and global change research [1].

Today, irrigated agriculture is one of the most significant contributors of water consumption [2], necessitating modeling water exchange between land surface and atmosphere [3], managing water resources [4] and analyzing the variability of irrigation water requirements and supply [5]. As a result, estimations of water demand do not consider the spatial variability in irrigation practices and do not reflect the characteristics of irrigation techniques [6].

Irrigation is currently used in most new olive orchards, which suggests that the percentage of irrigated olive trees is very high throughout the world and increasing [7]. Since water is the largest agricultural input in many cultivation systems in Mediterranean areas, with water availability being one of the main limiting factors for crop yield, the use of more efficient irrigation techniques in olive orchards has become essential [8]. Water Use Efficiency (WUE) is a term that was coined more than 100 years ago [9] and it functions as an indicator of the balance between productivity and water availability. WUE is an essential parameter nowadays due to the great pressure of increasingly intense and frequent droughts associated with climate change effects on agricultural water availability and crop yields worldwide [10]. As WUE is a measurement of yield or biomass produced per unit of water [11], it is therefore particularly useful when trying to compare the efficiency of different irrigation systems. In this context, many research projects demonstrate that productivity could be increased with no change in the rate of water use resulting in greater WUE [12–14]. For the olive grove, Martínez and Reca (2014) [8] compared Subsurface-Drip Irrigation (SDI) and surface Drip Irrigation (DI). Both yield and WUE for SDI outperformed DI, with a water savings of up to 20% for the former. As such, evaluating crop productivity and means of irrigation in relationship to WUE would allow a better management of water resources [10].

Remote sensing has been demonstrated to be an effective tool for locating, mapping and monitoring irrigation techniques by providing data in several regions of the electromagnetic spectrum and with a variety of spatial and temporal resolutions to assess crop growth, maturity and yield [15–18]. Images are obtained remotely via a broad range of sensors on-board three main types of platforms: satellite, manned aerial and unmanned [19]. Each of these approaches have pros and cons that involve economic, operational, and technological factors [19]. Satellite imagery covers extensive areas, and some Earth observation programs provide free low spatial-resolution datasets, e.g., Moderate Resolution Imaging Spectroradiometer (MODIS), or medium resolution datasets, e.g., Sentinel-1 and -2 from the Copernicus program [20,21]. The effectiveness of satellite imagery applied to arable crops, forests and extensive plantations has been demonstrated in many studies [1,21]. However, satellite imagery may suffer from cloud cover and constrain image timing for specific phenological characteristics due to the limits of temporal resolution [22]. Remote sensing becomes more challenging when considering crops with discontinuous layouts, such as olive trees, vineyards or orchards [23]. The presence of inter-row paths may deeply affect the overall computation of spectral indices, leading to an inadequate assessment of crop status [23]. On the other hand, manned aircraft surveys offer more operational flexibility, providing spatial resolutions in the range of centimeters, but comes with high operational and logistic costs [24,25], making it difficult to perform frequent flights in phenological studies [26]. UAV platforms offer greater flexibility still [27], allowing for the possibility to differentiate pure vegetation pixels in images over woody crops. However, UAV platforms are limited both in regard to payload and flight time [28]. Despite these limitations, UAV platforms have been shown to be very useful tools in the mapping of irrigated areas when working in concordance with traditional platforms [29]. In the context of this research, UAV platforms have been used in post-classification correction of traditional platforms to identify anomalies in the mapped irrigated plots and improve classification accuracy [29].

The UAV payload included a variety of sensors, including RGB, multispectral, hyperspectral, thermal and LiDAR. Multi- and hyper-spectral sensors have been successfully used in many applications which require accurate spectral information [30,31]. The main difference between both type of sensors is based on the number of spectral bands. While multispectral images generally range from 4 to 12 spectral bands that are represented in each pixel, hyperspectral images consist of hundreds of spectral bands arranged in a very narrow bandwidth [32]. The high spectral resolution in hyperspectral images allow the detection of spectral details that can be imperceptible in multispectral images due to their discrete spectral nature [32]. UAV-based hyperspectral imaging in agriculture has been successfully used in chlorophyll [33], biomass [34], nitrogen [35] or water [36] content estimation; the detection of diseases [37]; weed classification [38]; the evaluation and classification of crop water status [39]; etc. Therefore, hyperspectral remote sensing technologies have improved our capability for understanding the processes of biophysical and biochemical properties of vegetation [40].

As a result of the high number of spectral bands, many of them are highly correlated and therefore a dimension reduction or wavelength selection method is essential to apply in pre-processing of the hyperspectral image to improve its usability [41]. These selection methods can be grouped in three categories: (i) wave band features [42,43]; (ii) spectral position features [44,45]; and (iii) vegetation indices [45,46]. These methods are performed through a variety of techniques such as Principal Components Analysis (PCA) [47,48], Minimum Noise Fraction [49,50], Singular Value Decomposition [51] or Partial Least Square (PLS) [52,53] among others. As a result, these techniques reduce the data size by selecting those wavelengths sensitive to the object of interest [54]. Specifically, PLS regression is a nonparametric and supervised technique particularly useful to achieve the "large p -small n" problem in high-dimensional datasets [55]. This technique combines features from PCA and multiple regression to predict dependent variables from a set of orthogonal factors (latent variables) extracted from predictors with the best predictive power [56]. To achieve this, a simultaneous decomposition of predictors and dependent variables are performed with the constraint that these components explain the covariance as much as possible. Afterwards, a regression step is performed where the decomposition of predictors is used to predict dependent variables [56].

Previous studies have identified wavelengths that are sensitive to crop properties such as chlorophyll content, nitrogen status and water content or estimation of biomass [57–60]. PLS methods have proven to be very versatile for multivariate data analysis in applications related to bioinformatics [61,62] or chemometrics [63,64], as well as remote sensing [65,66]. Initially, in these studies, PLS was not implemented to select variables, since the objective was to find a relevant linear subspace of the explanatory variables, but, eventually, several PLS selection methods for variable selection were finally proposed [67]. These methods can be categorized in three types: filter, wrapper and embedded methods. Filter techniques evaluate the relevance of the characteristics by only looking at the intrinsic properties of the data. In most cases, a feature relevance score is calculated, and low scoring features are removed. Subsequently, this subset of characteristics is presented as input to the classification algorithm [67]. These methods require some type of filter measure (loading weights, regression coefficients or importance of the variable in projection) that represents the response relationship with the respective variable, and, for this, a threshold is required to classify the variables as selected or not [55]. For Loading Weights Method (LW-PLS), the peaks or valleys with the maximum absolute load weights from the first major factor to the optimal principal factor are selected as sensible wavelengths [68]. For the Regression Coefficient Method (RC-PLS), the sensitive wavelengths are generally selected according to the regression coefficient of the optimal PLS model [67]. In general, the peaks or bands are selected as the sensible wavelength or waveband when the absolute value of the regression coefficient is greater than the threshold [69]. On the other hand, the basis of the Variable Importance in Projection Method (VIP-PLS) is to accumulate the importance of each variable, this being reflected by the load weights from each component [70]. While filter techniques address the problem of finding a good subset of features regardless of the model selection step, wrapper methods incorporate the model hypothesis search into the feature subset search [67]. The methods are mainly distinguished by the choice of the underlying filter method and how the wrapper is implemented, and they are primarily based on procedures that iterate between model fitting and variable selection [55]. These procedures, for example, a Genetic Algorithm integrated with the PLS regression Method (GA-PLS), combine the advantage of GA and PLS and return a vector of variable numbers that corresponds to the model that has the lowest prediction error [71]. For Uninformative Variable Elimination PLS (UVE-PLS), artificial noise variables are added to the predictor set before the PLS model is fitted and all original variables that are less important than the artificial noise variables are removed before the procedure is repeated until a stop criterion is reached [72]. In the general procedure for Backward Variable Elimination PLS (BVE-PLS), variables are ordered first with respect to some measure of importance, and one of the filter measures described above is generally used [73]. Secondly, a threshold is used to eliminate a subset of the least informative variables, and then a model is fitted again to the remaining variables and performance is measured. The procedure is repeated until the maximum performance of the model is achieved. Other wrapper methods are Subwindow Permutation Analysis coupled with PLS (SwPA-PLS), which provides the influence of each variable without considering the influence of the rest of the variables [74]; Iterative Predictive Weighting PLS (IPW-PLS), which is an iterative elimination procedure where a measure of predictor importance is computed after fitting a PLS model [75]; Regularized Elimination Procedure in PLS (REP-PLS), where a stability-based variable selection procedure is adopted [67]; Backward [76] and Forward [77] Interval PLS (BiPLS and FiPLS, respectively), where the dataset is divided into a given number of intervals, and the PLS models are then calculated with each interval left in a sequence, giving the first omitted interval the worst performance model with respect to the mean squared error of cross-validation (MSECV); and Competitive Adaptive Reweighted Sampling (CARS-PLS), which is a function variable selection method that combines Monte Carlo sampling with the PLS regression coefficient [78]. For embedded methods, the search for an optimal subset of features is built into the classifier construction and can be seen as a search in the combined space of feature subsets and hypotheses [79]. Similar to wrapper approaches, embedded approaches are thus specific to a given learning algorithm. One of these methods is Sparse-PLS (S-PLS), which is a version of PLS that aims to combine selection and modeling in a one-step procedure [80].

On the other hand, machine learning algorithms, including Linear Discriminant Analysis (LDA) and K-Nearest Neighbors (KNN), are powerful tools for analyzing hyperspectral information since they can process a large number of variables efficiently [81,82]. LDA is a subspace technique that seeks to find the maximum Fisher's ratio [83] while KNN is a non-parametric learning algorithm since there is no assumption for the underlying data distribution. KNN also utilizes lazy learning, meaning that it does not need any training data point for the generation of the model [84]. Specifically, these machine learning algorithms have been widely used in the remote sensing field for agricultural applications [37,85]. For instance, Suarez et al. [85] estimated phenoxy herbicide dosage in cotton crops through the analysis of hyperspectral data with LDA. In addition, Bohnenkamp et al. [37] utilized KNN to detect yellow rust in wheat with the application of hyperspectral imaging technology. In general, these models have shown to be effectives for investigating agricultural features using hyperspectral imagery [86].

Many authors have compared selection PLS methods to analyze which offers better performance [69,87]. Likely, there is no best variable selection method due to the interaction between method and properties of the analyzed data [55]. As per the above discussion, this article focuses on the use of wavelength selection methods in UAV hyperspectral images to compare two irrigation systems commonly used in olive orchards, SDI and DI. For this purpose, 16 methods (13 based on PLS) were evaluated, and the quality of the results were assessed by two linear and nonlinear classification techniques. This paper is organized as follows. Section 2 describe the materials and methods used. Section 3 shows the results. Section 4 includes the final concluding remarks.

2. Materials and Methods

The workflow for classifying olive tree crowns according to the irrigation technique used is summarized in Figure 1. Two UAV flights were performed using RGB and hyperspectral sensors, respectively. The RGB UAV flight was used to geometrically define each of the olive trees while the hyperspectral UAV flight was used to characterize them radiometrically. Then, different PLS- and non-PLS-based methods were used to select the most significant wavelengths to classify the two irrigation techniques methods of the study area. Finally, with the selected wavelengths, two classifications were performed using LDA and KNN, evaluating the quality and efficiency of the results.

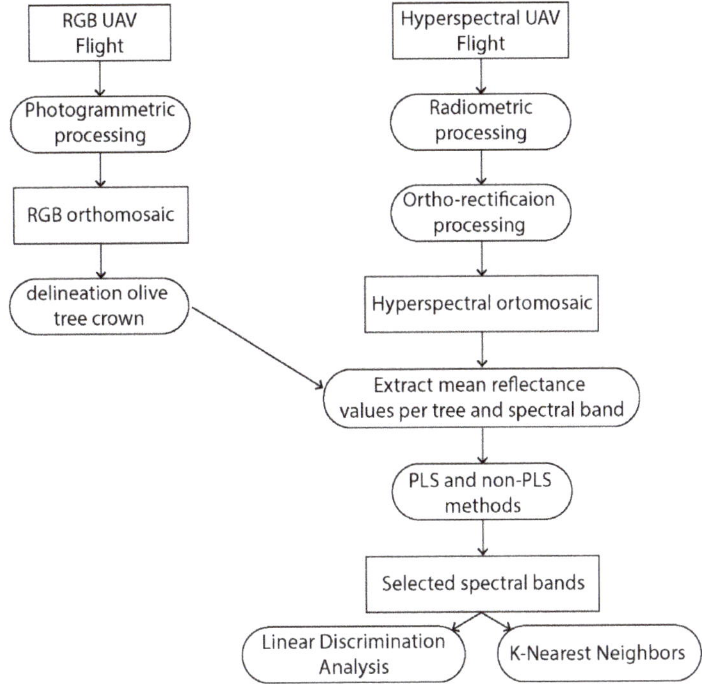

Figure 1. Flowchart used for the classification of irrigation techniques.

2.1. Study Area and UAV Flights

The study site was located in Marchena (Seville province, Spain) (37°24′52.7″ N, 5°27′44.9″ W, WGS84) (Figure 2a) in a 7-ha commercial orchard planted in 2015 with the olive cultivar Arbequina at a spacing of 3 m × 1.5 m (Figure 2b). The hedgerow orchard was adapted in equal parts to DI or SDI with carry self-compensating dripper pipes at 2.2 L/h spaced 0.5 m apart along the irrigation line and placed 0.5 m from the trunks and, in the case of SDI, buried 0.45 m deep. The trees were irrigated weekly with the seasonal water amount equivalent to 100% of ETc. Soil properties and agricultural practices were similar in both areas where the irrigation systems were implemented. The climate is Mediterranean with an average annual temperature and precipitation (concentrated mostly in late fall and winter) of 18.8 °C and 544 mm, respectively. A representative area including 413 olive trees (200 olives trees irrigated with DI and 213 with SDI) was used for data analyses.

2.2. UAV Flights and Processing

Two UAV flights, RGB and hyperspectral, were performed on 22 May 2020 at solar noon to take advantage of the position of the sun and minimize shadows in the images acquired. The RGB UAV flight was used to precisely delineate each of the olive trees in the study area while the hyperspectral UAV flight characterized each of them spectrally. A DJI Mavic Pro 2 (SZ DJI Technology Co., Shenzhen, China) was used to perform the RGB flight. As the payload, a Hasselblad L1D-20c (Hasselblad Group, Göteborg, Sweden) was used, which provides an image of 13.2 mm × 8.8 mm, a focal length of 10.3 mm and an image size of 5472 × 3648 pixels. This UAV was flown at an altitude of 65 m above ground level (AGL) and forward and side-lap were 80% and 70%, respectively. Ground Sample Distance (GSD) was 1.6 cm taking into account the characteristics of the sensor. For the hyperspectral flight, a DJI Matrice 600 Pro (SZ DJI Technology Co., Shenzhen, China) was used. This UAV platform was also equipped with a Nano Hyperspec sensor (Headwall Photonics Inc., Boston, MA, USA) with 270 spectral bands

(6-nm FWHM) both in the VNIR spectral range (400–1000 nm) with a sampling of 2.2 nm. Being a push-broom sensor, imagery is collected, line by line, at 12-bit radiometric resolution along the flight path where each line of pixels comprises 640 spatial pixels. The hyperspectral sensor was mounted on a Ronin-MX gimbal system (SZ DJI Technology Co., Shenzhen, China) to minimize external disturbances such as roll, pitch and yaw oscillations. This UAV platform was flown at an altitude of 100 m AGL, GSD being equal to 6 cm taking into account the characteristic of the sensor.

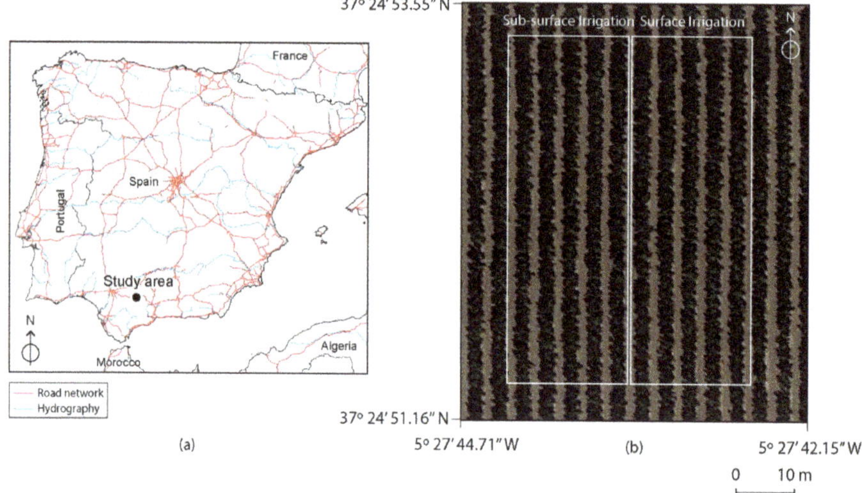

Figure 2. Study area: (**a**) Location; and (**b**) olive tree plantation details.

Prior to the UAV flights, five artificial targets were placed in the study area as Ground Control Points (GCPs), one in each corner and the other in the center. Each GCP was measured with the stop-and-go technique through relative positioning by means of the NTRIP protocol (The Radio Technical Commission for Maritime Services, RTCM, for Networked Transfer via Internet Protocol) using two GNSS (Global Navigation Satellite System) receivers. One of the receivers was a reference station for the GNSS Red Andaluza de Posicionamiento (RAP) network from the Institute of Statistics and Cartography of Andalusia, Spain, and the other, a Leica GS15 GNSS (Leica Geosystems AG, Heerbrugg, Switzerland), functioned as the rover receiver. In addition, a known reflectance value calibration tarp was placed in the center of the plot to subsequently correct the spectral cubes radiometrically. Before the hyperspectral UAV flight, the sensor capture mode was configured to set up the number of frames per second in accordance with the flight speed and AGL flight height. Likewise, the exposure level was established through the determination of the reflectance curve of a 90% target in order to avoid saturating the sensor in the recording of the spectral cubes. In addition, black body spectral information was taken by closing the aperture to later convert the digital levels to radiance values.

To produce an RGB orthomosaic from the RGB UAV flight, photogrammetric processing was divided into four stages: aerial triangulation, Digital Surface Model (DSM) generation, rectification of individual images and orthomosaicking. Aerial triangulation determined the individual external orientation of each image of the photogrammetric block. Afterwards, a dense point cloud was generated using Structure from Motion (SfM) techniques [88] to create a DSM. Finally, individual images were rectified and mosaicked to generate an RGB UAV orthomosaic of the study area. This methodology has been validated in previous research projects [19,89,90] and was performed using Pix4Dmapper (Pix4D S.A., Prilly, Switzerland). The hyperspectral data processing was divided into three stages [91]. Firstly, each hyperspectral cube was converted from digital number to radiance values, using dark reference captured prior to the UAV flight. Secondly, radiance values were converted to reflectance values using the values from the calibration tarp. Finally, all the reflectance hyperspectral cubes were orthorectified using data from the GNSS receivers, the Inertial Measurement Unit (IMU) and the Shuttle Radar Topography Mission (SRTM) Digital Elevation Model [92]. The accuracy and spatial resolution of the SRTM model was adequate due to the smooth relief of the study area. This processing was performed using SpectralView software (Headwall Photonics Inc., Boston, MA, USA).

Due to the difficulty in segmenting individual trees, an RGB orthomosaic was used to manually digitize polygons of all the olive trees, using the QGIS desktop Geographic Information System. As an example, a partial view of the hedgerow in the RGB and hyperspectral orthomosaic is shown in Figure 3. Then, for each olive tree, the mean spectral reflectance value for each of the 270 spectral bands was calculated, generating a spectral curve at the canopy level. This step was automated using a script developed in the Python programming language [93].

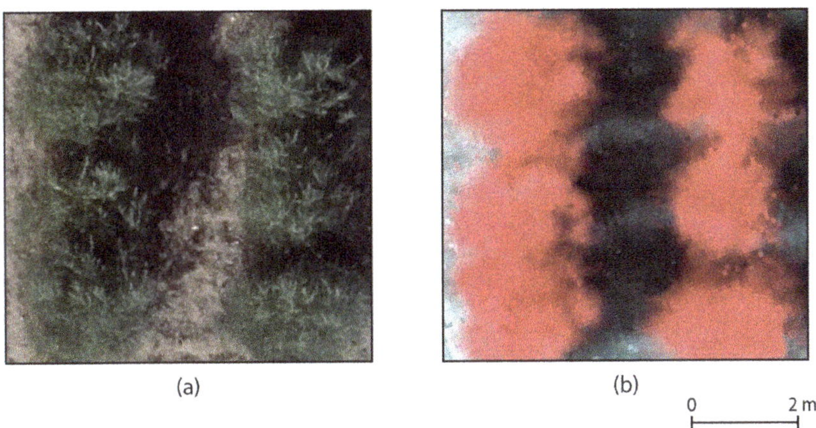

Figure 3. Partial view of (**a**) RGB and (**b**) hyperspectral orthomosaic of the study area.

2.3. Wavelength Selection Methods Used and Evaluation

Although there are other methods, this project focused on applying those methods that use the output of a PLS algorithm to identify a subset of important variables. As such, the variables analyzed in this project were the wavelengths registered by the hyperspectral sensor. The PLS selection methods used in this study are grouped by category (Table 1). More information on the PLS methods used in this study can be found in the work by Mehmood et al. [67].

Other wavelength selection methods, such as the Least Absolute Shrinkage and Selection Operator (Lasso) and Boruta algorithms, were also included in this study [94,95]. Lasso can handle ill-posed problems (i.e., a large number of correlated variables compared to sample size) and works by penalizing the magnitude of the characteristic coefficients along with minimizing the error between the predicted and actual observations [96]. On the other hand, the Boruta algorithm was developed to identify all relevant variables within a classification framework [97]. This method searches for relevant characteristics by comparing the importance of the original attributes with the importance that can be obtained at random, estimated using their permuted copies, and progressively eliminating irrelevant characteristics to stabilize the test. First, the dataset is duplicated, and the values are shuffled in each column; these are called shadow characteristics. Then, a classifier is trained on the dataset to extract the importance of each characteristic. Random Forest is one of the most widely used classifiers for this purpose [97]. As a final step, a combined analysis was performed considering the individual results from all the PLS, Lasso and Boruta methods used, which has been termed the All-together method.

Table 1. Evaluated wavelength selection PLS methods.

Category	PLS Method	Reference
Filter	Loading Weights (LW-PLS)	[68]
	Regression Coefficient (RC-PLS)	[69]
	Variable Importance in Projection (VIP-PLS)	[70]
Wrapper	Genetic Algorithm (GA-PLS)	[71]
	Uninformative Variable Elimination (UVE-PLS)	[72]
	Backward Variable Elimination (BVE-PLS)	[73]
	Subwindow Permutation Analysis (SwPA-PLS)	[74]
	Iterative Predictive Weighting (IPW-PLS)	[75]
	Regularized Elimination Procedure (REP-PLS)	[67]
	Backward Interval (BiPLS)	[76]
	Forward Interval (FiPLS)	[77]
	Competitive Adaptive Reweighted Sampling (CARS-PLS)	[78]
Embedded	Sparse (S-PLS)	[80]

2.4. Evaluation of Wavelength Selection Methods

LDA and KNN were applied to classify irrigation techniques used in each olive tree. Of the total number of olive trees, 75% were used in the calibration phase (explained below) and 25% for prediction (Table 2). Before applying LDA or KNN classification, the dataset was divided into two smaller datasets that were used for calibration and prediction purposes. The calibration subset of data was used to estimate the parameters of the classifier model and the prediction subset of data was used to check the results of the model. Calibration and prediction were performed following an iterative process where subsets of data changed per iteration.

Table 2. Number of samples per irrigation technique for calibration and prediction set.

	Sub-Surface Irrigation	Surface Irrigation
Calibration set	160	150
Prediction set	53	50
Total	213	200

Overall accuracy (OA) of LDA and KNN results were calculated by summing the number of correctly classified olive trees and dividing by the total number of trees. Moreover, the accuracy of each irrigation technique was evaluated [98]. In addition, efficiency of methods applied were calculated in the prediction stage. According to Xia et al. [87], the efficiency of a wavelength selection method is based on the prediction rate and the number of variables, being calculated as follows:

$$E = \frac{\left(D_s - D_f\right) \times \left(N_f - N_s\right)}{N_f} \times 100 \quad (1)$$

where E is the efficiency of the wavelength selection method evaluated, D_s is the OA obtained by a PLS method, D_f is the OA obtained using all wavelengths registered by the UAV hyperspectral sensor, N_f is the total number of wavelengths and N_s is the number of wavelengths selected by the method used. If E is higher than or equal to 0.5, the selection method is highly efficient. On the other hand, if E ranges from −0.5 to 0.5 the method is shown to be efficient, except if E is equal to 0 and $\left(N_f - N_s\right)/N_f$ is greater than or equal to 0.8, indicating the method to be highly efficient. Finally, if E is less than or equal to −0.5 the method is of low efficiency.

2.5. Software

LW-PLS, RC-PLS, VIP-PLS, AG-PLS, UVE-PLS, BVE-PLS, SwPA-PLS, IPW-PLS, REP-PLS, BiPLS, FiPLS, CARS-PLS, S-PLS, Lasso and Boruta were conducted within the R software environment [99] using RStudio [100], with the packages shown in Table 3. For the evaluation of the identification, linear and nonlinear techniques were established through LDA and KNN, respectively. The implementation of these algorithms was performed in Python [93] using the Scikit-learn library [84].

Table 3. R packages used for wavelength selection methods.

PLS Method	R Packages	Reference
Loading Weights (LW-PLS)	plsVarSel, pls	[67,101]
Regression Coefficient (RC-PLS)	plsVarSel, pls, threshr	[67,101,102]
Variable Importance in Projection (VIP-PLS)	plsVarSel, pls	[67,101]
Genetic Algorithm (GA-PLS)	plsVarSel	[67]
Uninformative Variable Elimination (UVE-PLS)	plsVarSel	[67]
Backward Variable Elimination (BVE-PLS)	plsVarSel	[67]
Subwindow Permutation Analysis (SwPA-PLS)	plsVarSel	[67]
Iterative Predictive Weighting (IPW-PLS)	plsVarSel	[67]
Regularized Elimination Procedure (REP-PLS)	plsVarSel	[67]
Backward Interval (BiPLS)	mdatools	[103]
Forward Interval (FiPLS)	mdatools	[103]
Competitive Adaptive Reweighted Sampling (CARS-PLS)	libPLSn	[104]
Sparse (S-PLS)	spls	[105]
Lasso	glmnet	[106]
Boruta	Boruta	[97]

3. Results and Discussion

3.1. Spectral Reflectance Data

The mean spectral reflectance curves of olive trees, being irrigated with SDI or DI systems, were similar to the results obtained in related studies [107] (Figure 4). No differences were observed in the 400–550 and 880–947 nm ranges. However, different magnitudes of spectral reflectance were found in the 550–880 nm range. Therefore, within that range, there are 165 possible wavelengths to differentiate between both irrigation techniques. Given this high number, different wavelength selection methods, detailed above, were applied to identify those that best classify both types of irrigation.

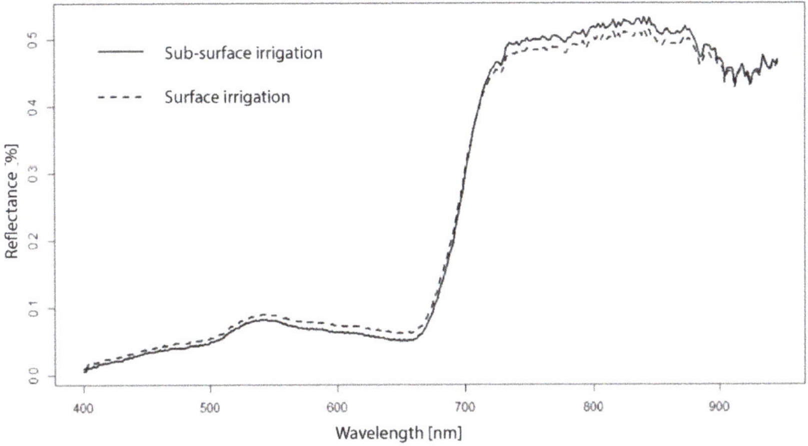

Figure 4. Average raw spectra reflectance curves of olive canopies irrigated with SDI or DI.

3.2. Wavelength Selection Results

The results of the application of the methods used, including the number of the selected wavelengths as well as their wavelengths, are shown in Table 4. Depending on the method employed, the number of wavelengths varied considerably. The method that selected the least number of wavelengths was CARS, followed by LW-PLS, neither exceeding five wavelengths, less than 1% of the original wavelengths registered. A set of methods (RC-PLS, VIP-PLS and IPW-PLS) selected a range of 10 wavelengths, equal to 3.6% of the original wavelength number, and others selected 17–77 (6.2–28.1%) wavelengths (GA-PLS, BVE-PLS, SwPA-PLS, REP_PLS, FiPLS, Lasso and Boruta). Finally, the S-PLS and BiPLS methods were the ones that selected the largest number of wavelengths, 192 (70%) and 265 (96.7%), respectively. As such, the reduction in the number of wavelengths of interest will be higher or lower depending on the method used.

Table 4. Wavelengths selected by different methods.

Method	Number of Wavelengths	Wavelengths [nm]
LW-PLS	5	882, 884, 890, 934, 942
RC-PLS	12	726, 728, 888, 904–908, 914, 920, 924, 928, 930–942
VIP-PLS	10	726, 888, 904, 906, 914, 924, 928, 936, 938, 942
GA-PLS	31	424, 428, 436, 442, 444, 458, 460, 522, 588, 612, 630, 640, 662, 698, 714, 716, 744, 758, 770, 780, 826, 846, 854, 860, 870, 878, 888, 912, 918, 920, 938
UVE-PLS	10	428, 696, 698, 700–704, 734, 792, 812, 856
BVE-PLS	69	686–734, 746, 776, 790, 840, 846, 848, 858, 860, 864–870, 880–890, 894–946
SwPA-PLS	77	410, 414, 418, 424, 436, 450, 466, 476, 490, 494, 496, 500, 506, 508, 518, 530–538, 542, 558, 562, 568, 576–580, 584, 588, 596, 612, 622–630, 640, 646, 648, 660, 668, 690, 702–706, 728, 738, 746, 752, 756, 768, 774, 782, 786, 804, 812, 822, 828, 836, 838, 848, 850, 856, 858, 860, 862, 874, 878, 882, 888, 896, 900, 904, 910, 914, 926, 936, 940
IPW-PLS	12	710, 790, 832, 846, 888, 914, 920–924, 936–940
REP-PLS	31	726, 728, 882, 884, 888, 890, 894, 898, 902–946
BiPLS	265	400–604, 624–946
FiPLS	54	660–676, 696–730, 768–784, 858–874, 912–928
CARS	2	436, 790
S-PLS	192	560–714, 720–946
Lasso	17	404, 662, 698, 704, 788, 844, 846, 858, 886, 904, 912, 918–922, 934–938
Boruta	29	430, 456, 464, 474, 636, 638, 644, 646, 650–660, 698, 774, 776, 780, 792, 794, 810, 816, 818, 838, 866, 868, 914, 930, 936
All-together	18	790, 814, 846, 860, 866, 868, 870, 882, 888, 892, 898, 902, 904, 920, 928, 934, 936, 940

The total number of times a wavelength was selected by the total number of methods simultaneously employed is shown in Figure 5. Although there was a difference in the spectrum of SDI and DI in the wavelengths of around 600–650 nm (Figure 4), the frequency of selection by the selection methods employed in this range was low (Figure 5). Of all the methods applied, 18 wavelengths were selected by almost half the methods (All-together method; Table 4). All within the infrared region and for a total of nine times, the wavelength 920 nm was selected the most to predict the irrigation method used in this study. The utility of the All-together method was demonstrated for investigating agricultural features using hyperspectral imagery.

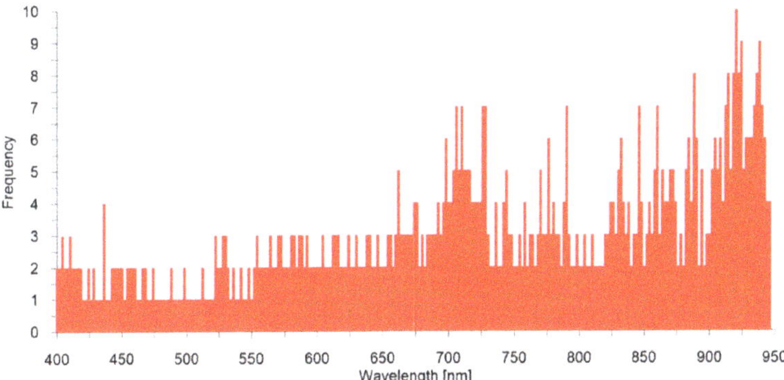

Figure 5. Number of times a wavelength has been selected when performing the All-together method.

3.3. *Quality and Efficiency of Classification with Each Selection Method*

The accuracy and efficiency results of the irrigation technique rating using LDA and KNN methods and their efficiency are shown in Table 5, while Figure 6 shows their efficiency. In general terms, LDA classifications showed better overall accuracy than KNN, in both calibration and prediction. Considering LDA, overall accuracy prediction (OAP) using all wavelengths was equal to 65%. Using only those selected wavelengths from a selection method, OAP improved, except in LW-PLS, the mean OPA being equal to 75%. The highest percentage of improvement was offered by GA-PLS, at 20%, while the lowest was offered by CARS, at 1%. In addition, ten of the methods evaluated improved the predictive quality of the irrigation technique by more than 10% (RC-PLS, GA-PLS, BVE-PLS, SwPA-PLS, IPW-PLS, BiPLS, S-PLS, Lasso, Boruta and All-together). On the other hand, the OAP applying KNN using all wavelengths was equal to 68.1% while just using the wavelengths selected by a PLS method generally resulted in a slightly lower OAP, with a mean value equal to 66.4%. Boruta was the method that showed the highest percentage of improvement, with an OAP of 74%. In total, seven of the methods offered an improvement in irrigation system prediction (GA-PLS, IPW-PLS, BiPLS, S-PLS, Lasso, Boruta and All-together). In addition, while LDA showed similar accuracy classifying both irrigation techniques, KNN offered worse results with the SDI technique.

Table 5. Overall accuracy (OA), Accuracy of Sub-Surface drip irrigation (A SDI), Accuracy of Surface drip irrigation (A DI) and Efficiency (E) results of Linear Discriminant Analysis (LDA) and K-Nearest Neighbors (KNN) using different selection methods.

	LDA				KNN			
Method	**OA (%)**	**A SDI (%)**	**A DI (%)**	**E**	**OA (%)**	**A SDI (%)**	**A DI (%)**	**E**
All bands	65.0	62.5	68.7	-	68.1	64.6	73.6	-
LW-PLS	62.5	57.5	70.7	−2.9	54.8	51.7	61.1	−13.1
RC-PLS	81.3	80.7	81.1	15.3	68.3	64.4	73.3	0.2
VIP-PLS	72.2	69.5	73.3	6.7	60.6	55.1	71.9	7.2
GA-PLS	85.2	83.6	87.6	17.7	69.2	65.4	73.0	1.0
UVE-PLS	66.5	61.9	72.8	1.0	63.5	59.9	70.1	−4.4
BVE-PLS	79.0	78.5	79.5	10.5	65.4	61.7	71.1	−2.0
SwPA-PLS	76.3	76.2	76.4	7.9	68.3	66.6	70.0	0.1
IPW-PLS	75.5	71.5	79.5	9.6	71.2	67.4	75.4	3.0
REP-PLS	69.3	67.9	71.3	3.5	59.6	54.5	71.1	−7.5
BiPLS	70.2	68.3	72.1	0.2	69.2	64.2	75.2	0.0
FiPLS	80.1	81.6	79.0	12.0	66.3	63.5	70.1	−1.4
CARS	66.0	60.1	76.1	1.0	63.5	60.1	67.9	−4.6
S-PLS	75.3	75.3	75.3	3.0	69.2	66.2	72.2	0.5
Lasso	84.0	86.4	82.3	17.8	71.2	72.3	72.2	2.9
Boruta	78.0	76.4	80.2	11.6	74.0	73.5	75.6	5.3
All-together	82.5	78.6	85.3	15.9	69.2	65.3	74.9	1.0

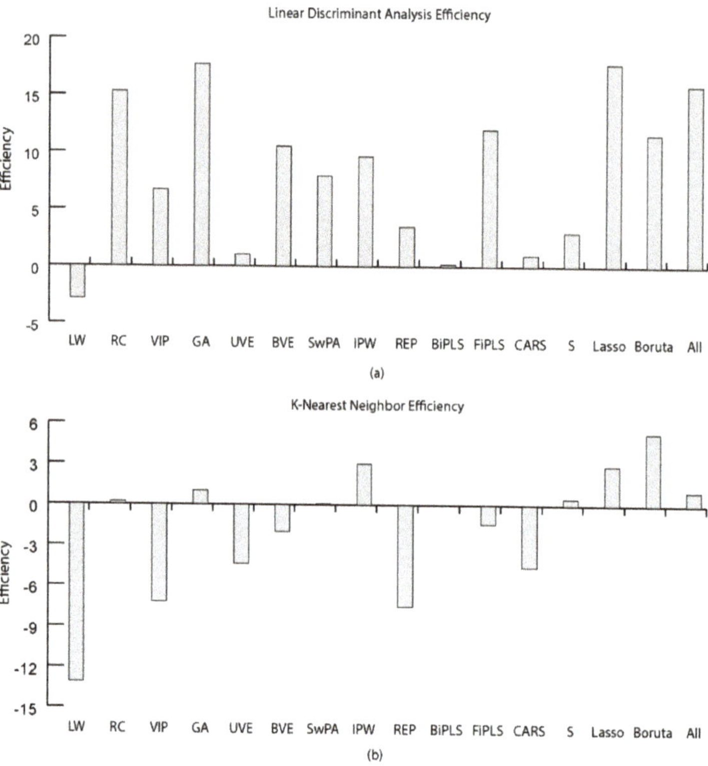

Figure 6. Efficiency of PLS method using (**a**) Linear Discriminant Analysis and (**b**) K-Nearest Neighbor.

While LDA offered better results than KNN, all the methods used were highly efficient except for LW-PLS (Figure 6). GA-PLS and Lasso scored the highest with an efficiency equal to 17.7 and 17.8, respectively (Figure 6a). In contrast, only six of the sixteen methods were highly efficient using KNN, Boruta being the most efficient with a value equal to 5.3 (Figure 6b). In addition, the irrigation classification maps of some of the selection methods assessed are shown in Figure 7. From a visual analysis, no spatial correlation was detected in the errors obtained as well as the presence of a higher concentration of errors in the perimeter of each type of irrigation.

Although combining bands when using hyperspectral data for olive trees is common [108], this study shows that the use of individual wavelengths could be an interesting and more accessible way to manage hyperspectral data. The results obtained from each selection method were different, showing methods that improved overall accuracy and efficiency in the classification of irrigation systems and methods that did not. This variation in the results obtained by selection methods were similar to those obtained by other authors [87]. In this study, according to the LDA and KNN results, GA-PLS, IPW-PLS, Lasso, Boruta and All-together methods were those which improved OAP while being highly efficient at the same time. In addition, RC-PLS and FiPLS methods showed a high overall accuracy using LDA. The GA-PLS method was also one of the most efficient, as seen by Xia et al. (2017) [87]. These authors found spectral differences between different samples of *Ophiopogon japonicus* from differing origins, using, in their case, an imaging spectrograph. In addition, although the Boruta method has not been widely used for the selection of variables in remote sensing or other disciplines, results obtained in this study show that it could be considered as a promising method for the selection of hyperspectral wavelengths. On the other hand, LW-PLS showed low efficiency and lower OAP. This can be explained because this method greatly reduces wavelengths and therefore eliminated useful information as indicated in other works [87]. The efficiency of RC-PLS, VIP-PLS, UVE-PLS, BVE-PLS, SwPA-PLS, REP-PLS, BiPLS, FiPLS, CARS-PLS and S-PLS varied depending on whether LDA or KNN was used. In general, LDA showed better classification results than KNN, which can be explained due to the fact that KNN needs to have large training data in order to achieve acceptable classification results, as indicated by Starzacher and Rinner (2008) [109].

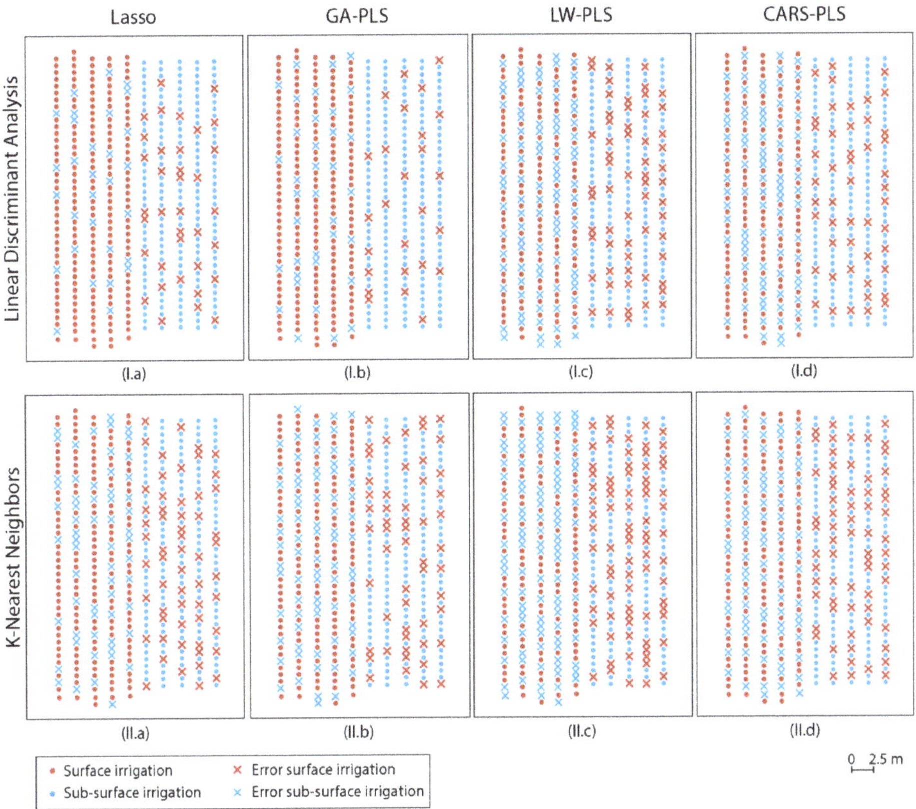

Figure 7. Classification results per classifier ((I) Linear Discriminant Analysis; and (II) K-Nearest Neighbors) and PLS method: (**a**) Lasso, (**b**) Genetic Algorithm, (**c**) Loading Weights; and (**d**) Competitive Adaptative Reweighted Sampling.

In general, these preliminary results show the need to analyze the relationships between wavelengths registered by a hyperspectral sensor and the object of study to optimize the number of wavelengths utilized. As a future line of research, the analysis of which combinations of wavelengths may be of interest to characterize irrigated areas in more detail is proposed. In addition, other features such as soil properties or olive cultivars along with other classifiers should be considered for improving the characterization of irrigation areas.

4. Conclusions

This study explored the use of UAV hyperspectral reflectance measurements of olive trees as a means for differentiating irrigation systems. Because of the high dimension and multicollinearity of the data, selection methods were found appropriate due to their capacity to extract useful wavelengths for analyzing categorial data, even when using individualized wavelengths. The results showed how the spectral response of olive trees is sensitive to the irrigation technique used, allowing improved information for the mapping of irrigated areas. Overall accuracy in the classification of irrigation systems in olives trees using LDA and KNN ranged from 54.8% and 85.2%. These variations showed the need to select the appropriate wavelength selection method. In addition, LDA offered more accurate results than KNN. In our study, GA-PLS, RC-PLS, Lasso, FiPLS, Boruta and All-together showed an overall accuracy of 75% or higher. They were all highly efficient methods and resulted in an improved classification.

The study has shown how the use of hyperspectral UAV data allows the irrigated areas of olive groves to be characterized in greater detail. This will generate information of interest for decision-making processes in the context of water use policies, enabling better understanding of irrigated olive groves and improving the management of water resources.

Author Contributions: Conceptualization, A.S.-R. and F.-J.M.-C.; methodology, A.S.-R. and F.-J.M.-C.; resources, F.-J.M.-C., A.G.-F. and J.E.M.-L.; writing—original draft preparation, A.S.-R. and F.-J.M.-C.; and writing—review and editing, A.S.-R. and F.-J.M.-C. All authors have read and agreed to the published version of the manuscript.

Funding: This research received no external funding.

Acknowledgments: The authors thank the support of the Higher Technical School of Agricultural and Forestry Engineering (Escuela Técnica Superior Ingeniería Agronómica y de Montes) of the University of Córdoba, Spain.

Conflicts of Interest: The authors declare no conflict of interest.

References

1. Ozdogan, M.; Yang, Y.; Allez, G.; Cervantes, C. Remote sensing of irrigated agriculture: Opportunities and challenges. *Remote Sens.* **2010**, *2*, 2274–2304. [CrossRef]
2. Cai, X.; Rosegrant, M.W. Global Water Demand and Supply Projections: Part 1. A Modeling Approach. *Water Int.* **2002**, *27*, 159–169. [CrossRef]
3. Wisser, D.; Frolking, S.; Douglas, E.M.; Fekete, B.M.; Vörösmarty, C.J.; Schumann, A.H. Global irrigation water demand: Variability and uncertainties arising from agricultural and climate data sets. *Geophys. Res. Lett.* **2008**, *35*, 1–5. [CrossRef]
4. Kueppers, L.M.; Snyder, M.A.; Sloan, L.C. Irrigation cooling effect: Regional climate forcing by land-use change. *Geophys. Res. Lett.* **2007**, *34*, 1–5. [CrossRef]
5. Droogers, P.; Aerts, J. Adaptation strategies to climate change and climate variability: A comparative study between seven contrasting river basins. *Phys. Chem. Earth* **2005**, *30*, 339–346. [CrossRef]
6. Thenkabail, P.S.; Dheeravath, V.; Biradar, C.M.; Gangalakunta, O.R.P.; Noojipady, P.; Gurappa, C.; Velpuri, M.; Gumma, M.; Li, Y. Irrigated area maps and statistics of India using remote sensing and national statistics. *Remote Sens.* **2009**, *1*, 50–67. [CrossRef]
7. Escobar, R.F.; de la Rosa, R.; Leon, L.; Gomez, J.A.; Testi, F.; Orgaz, M.; Gil-Ribes, J.A.; Quesada-Moraga, E.; Trapero, A. Evolution and sustainability of the olive production systems. In *Present and Future of the Mediterranean Olive Sector*; CIHEAM/IOC: Zaragoza, Spain, 2013; Volume 106, pp. 11–41. ISBN 2-85352-512-0.
8. Martínez, J.; Reca, J. Water use efficiency of surface drip irrigation versus an alternative subsurface drip irrigation method. *J. Irrig. Drain. Eng.* **2014**, *140*, 1–9. [CrossRef]
9. Briggs, L.J.; Shantz, H.L. The water requirement of plants. In *Bureau of Plant Industry Bulletin*; Wiley: Akron, OH, USA, 1913; p. 96.
10. Yu, L.; Gao, X.; Zhao, X. Global synthesis of the impact of droughts on crops' water-use efficiency (WUE): Towards both high WUE and productivity. *Agric. Syst.* **2020**, *177*, 102723. [CrossRef]
11. Tolk, J.; Howell, T.; Evett, S. Effect of mulch, irrigation, and soil type on water use and yield of maize. *Soil Tillage Res.* **1999**, *50*, 137–147. [CrossRef]
12. Basso, B.; Ritchie, J.T. Evapotranspiration in High-Yielding Maize and under Increased Vapor Pressure Deficit in the US Midwest. *Agric. Environ. Lett.* **2018**, *3*, 170039. [CrossRef]
13. Bota, J.; Tomás, M.; Flexas, J.; Medrano, H.; Escalona, J.M. Differences among grapevine cultivars in their stomatal behavior and water use efficiency under progressive water stress. *Agric. Water Manag.* **2016**, *164*, 91–99. [CrossRef]
14. Michelon, N.; Pennisi, G.; Myint, N.O.; Orsini, F.; Gianquinto, G. Strategies for improved Water Use Efficiency (WUE) of field-grown lettuce (*Lactuca sativa* L.) under a semi-arid climate. *Agronomy* **2020**, *10*, 668. [CrossRef]

15. Thenkabail, P.S.; Schull, M.; Turral, H. Ganges and Indus river basin land use/land cover (LULC) and irrigated area mapping using continuous streams of MODIS data. *Remote Sens. Environ.* **2005**, *95*, 317–341. [CrossRef]
16. Toureiro, C.; Serralheiro, R.; Shahidian, S.; Sousa, A. Irrigation management with remote sensing: Evaluating irrigation requirement for maize under Mediterranean climate condition. *Agric. Water Manag.* **2017**, *184*, 211–220. [CrossRef]
17. Gao, Q.; Zribi, M.; Escorihuela, M.J.; Baghdadi, N.; Segui, P.Q. Irrigation mapping using Sentinel-1 time series at field scale. *Remote Sens.* **2018**, *10*, 1495. [CrossRef]
18. Jalilvand, E.; Tajrishy, M.; Hashemi, S.A.G.Z.; Brocca, L. Quantification of irrigation water using remote sensing of soil moisture in a semi-arid region. *Remote Sens. Environ.* **2019**, *231*, 111226. [CrossRef]
19. Mesas-Carrascosa, F.-J.; Torres-Sánchez, J.; Clavero-Rumbao, I.; García-Ferrer, A.; Peña, J.-M.; Borra-Serrano, I.; López-Granados, F. Assessing Optimal Flight Parameters for Generating Accurate Multispectral Orthomosaicks by UAV to Support Site-Specific Crop Management. *Remote Sens.* **2015**, *7*, 12793–12814. [CrossRef]
20. Pageot, Y.; Baup, F.; Inglada, J.; Baghdadi, N.; Demarez, V. Detection of Irrigated and Rainfed Crops in Temperate Areas Using Sentinel-1 and Sentinel-2 Time Series. *Remote Sens.* **2020**, *12*, 3044. [CrossRef]
21. Velpuri, N.M.; Thenkabail, P.S.; Gumma, M.K.; Biradar, C.; Dheeravath, V.; Noojipady, P.; Yuanjie, L. Influence of resolution in irrigated area mapping and area estimation. *Photogramm. Eng. Remote Sens.* **2009**, *75*, 1383–1395. [CrossRef]
22. Matese, A.; Toscano, P.; Di Gennaro, S.F.; Genesio, L.; Vaccari, F.P.; Primicerio, J.; Belli, C.; Zaldei, A.; Bianconi, R.; Gioli, B. Intercomparison of UAV, aircraft and satellite remote sensing platforms for precision viticulture. *Remote Sens.* **2015**, *7*, 2971–2990. [CrossRef]
23. Eastman, J.R.; Sangermano, F.; Machado, E.A.; Rogan, J.; Anyamba, A. Global trends in seasonality of Normalized Difference Vegetation Index (NDVI), 1982–2011. *Remote Sens.* **2013**, *5*, 4799–4818. [CrossRef]
24. Berni, J.A.J.; Zarco-Tejada, P.J.; Suárez, L.; Fereres, E. Thermal and Narrowband Multispectral Remote Sensing for Vegetation Monitoring from an Unmanned Aerial Vehicle Improved Evapotranspiration using Unmanned Aerial Vehicles View project High throughput and remote trait measurement View project Thermal and Nar. *IEEE Trans. Geosci. Remote Sens.* **2009**, *47*, 722–738. [CrossRef]
25. Anderson, K.; Gaston, K.J. Lightweight unmanned aerial vehicles will revolutionize spatial ecology. *Front. Ecol. Environ.* **2013**, *11*, 138–146. [CrossRef]
26. Berra, E.F.; Gaulton, R.; Barr, S. Assessing spring phenology of a temperate woodland: A multiscale comparison of ground, unmanned aerial vehicle and Landsat satellite observations. *Remote Sens. Environ.* **2019**, *223*, 229–242. [CrossRef]
27. Jay, S.; Baret, F.; Dutartre, D.; Malatesta, G.; Héno, S.; Comar, A.; Weiss, M.; Maupas, F. Exploiting the centimeter resolution of UAV multispectral imagery to improve remote-sensing estimates of canopy structure and biochemistry in sugar beet crops. *Remote Sens. Environ.* **2019**, *231*, 110898. [CrossRef]
28. Radoglou-Grammatikis, P.; Sarigiannidis, P.; Lagkas, T.; Moscholios, I. A compilation of UAV applications for precision agriculture. *Comput. Netw.* **2020**, *172*, 107148. [CrossRef]
29. Nhamo, L.; van Dijk, R.; Magidi, J.; Wiberg, D.; Tshikolomo, K. Improving the accuracy of remotely sensed irrigated areas using post-classification enhancement through UAV capability. *Remote Sens.* **2018**, *10*, 712. [CrossRef]
30. Qin, J.; Chao, K.; Kim, M.S.; Lu, R.; Burks, T.F. Hyperspectral and multispectral imaging for evaluating food safety and quality. *J. Food Eng.* **2013**, *118*, 157–171. [CrossRef]
31. Park, B.; Lu, R. (Eds.) *Hyperspectral Imaging Technology in Food and Agriculture*; Springer US: New York, NY, USA, 2015; ISBN 9781493928354.
32. Adão, T.; Hruška, J.; Pádua, L.; Bessa, J.; Peres, E.; Morais, R.; Sousa, J.J. Hyperspectral imaging: A review on UAV-based sensors, data processing and applications for agriculture and forestry. *Remote Sens.* **2017**, *9*, 1110. [CrossRef]
33. Yu, F.H.; Xu, T.Y.; Du, W.; Ma, H.; Zhang, G.S.; Chen, C.L. Radiative transfer models (RTMs) for field phenotyping inversion of rice based on UAV hyperspectral remote sensing. *Int. J. Agric. Biol. Eng.* **2017**, *10*, 150–157. [CrossRef]
34. Yue, J.; Yang, G.; Li, C.; Li, Z.; Wang, Y.; Feng, H.; Xu, B. Estimation of Winter Wheat Above-Ground Biomass Using Unmanned Aerial Vehicle-Based Snapshot Hyperspectral Sensor and Crop Height Improved Models. *Remote Sens.* **2017**, *9*, 708. [CrossRef]

35. Akhtman, Y.; Golubeva, E.I.; Tutubalina, O.V.; Zimin, M. Application of hyperspectural images and ground data for precision farming. *Geogr. Environ. Sustain.* **2017**, *10*, 117–128. [CrossRef]
36. Bauer, M.E.; Daughtry, C.S.T.; Vanderbilt, V.C. Spectral-agronomic relationships of corn, soybean and wheat canopies. In Proceedings of the Signatures Spectrales D'objets En Teledetection, Avignon, France, 18–22 January 1981; pp. 261–272.
37. Bohnenkamp, D.; Behmann, J.; Mahlein, A.K. In-field detection of Yellow Rust in Wheat on the Ground Canopy and UAV Scale. *Remote Sens.* **2019**, *11*, 2495. [CrossRef]
38. Scherrer, B.; Sheppard, J.; Jha, P.; Shaw, J.A. Hyperspectral imaging and neural networks to classify herbicide-resistant weeds. *J. Appl. Remote Sens.* **2019**, *13*, 044516. [CrossRef]
39. Rinaldi, M.; Castrignanò, A.; De Benedetto, D.; Sollitto, D.; Ruggieri, S.; Garofalo, P.; Santoro, F.; Figorito, B.; Gualano, S.; Tamborrino, R. Discrimination of tomato plants under different irrigation regimes: Analysis of hyperspectral sensor data. *Environmetrics* **2015**, *26*, 77–88. [CrossRef]
40. Renzullo, L.J.; Blanchfield, A.L.; Powell, K.S. A method of wavelength selection and spectral discrimination of hyperspectral reflectance spectrometry. *IEEE Trans. Geosci. Remote Sens.* **2006**, *44*, 1986–1994. [CrossRef]
41. Lu, B.; Dao, P.D.; Liu, J.; He, Y.; Shang, J. Recent Advances of Hyperspectral Imaging Technology and Applications in Agriculture. *Remote Sens.* **2020**, *12*, 2659. [CrossRef]
42. Thenkabail, P.S.; Enclona, E.A.; Ashton, M.S.; Van Der Meer, B. Accuracy assessments of hyperspectral waveband performance for vegetation analysis applications. *Remote Sens. Environ.* **2004**, *91*, 354–376. [CrossRef]
43. Becker, B.L.; Lusch, D.P.; Qi, J. Identifying optimal spectral bands from in situ measurements of Great Lakes coastal wetlands using second-derivative analysis. *Remote Sens. Environ.* **2005**, *97*, 238–248. [CrossRef]
44. Pu, R.; Gong, P.; Biging, G.S.; Larrieu, M.R. Extraction of red edge optical parameters from hyperion data for estimation of forest leaf area index. *IEEE Trans. Geosci. Remote Sens.* **2003**, *41*, 916–921. [CrossRef]
45. Darvishzadeh, R.; Atzberger, C.; Skidmore, A.K.; Abkar, A.A. Leaf Area Index derivation from hyperspectral vegetation indicesand the red edge position. *Int. J. Remote Sens.* **2009**, *30*, 6199–6218. [CrossRef]
46. Haboudane, D.; Miller, J.R.; Pattey, E.; Zarco-Tejada, P.J.; Strachan, I.B. Hyperspectral vegetation indices and novel algorithms for predicting green LAI of crop canopies: Modeling and validation in the context of precision agriculture. *Remote Sens. Environ.* **2004**, *90*, 337–352. [CrossRef]
47. Farrell, M.D.; Mersereau, R.M. On the impact of PCA dimension reduction for hyperspectral detection of difficult targets. *IEEE Geosci. Remote Sens. Lett.* **2005**, *2*, 192–195. [CrossRef]
48. Koonsanit, K.; Jaruskulchai, C.; Eiumnoh, A. Band Selection for Dimension Reduction in Hyper Spectral Image Using Integrated InformationGain and Principal Components Analysis Technique. *Int. J. Mach. Learn. Comput.* **2012**, *2*, 248–251. [CrossRef]
49. Burger, J.; Gowen, A. Data handling in hyperspectral image analysis. *Chemom. Intell. Lab. Syst.* **2011**, *108*, 13–22. [CrossRef]
50. Gao, L.; Zhao, B.; Jia, X.; Liao, W.; Zhang, B. Optimized kernel minimum noise fraction transformation for hyperspectral image classification. *Remote Sens.* **2017**, *9*, 548. [CrossRef]
51. Menon, V.; Du, Q.; Fowler, J.E. Fast SVD with Random Hadamard Projection for Hyperspectral Dimensionality Reduction. *IEEE Geosci. Remote Sens. Lett.* **2016**, *13*, 1275–1279. [CrossRef]
52. Fordellone, M.; Bellincontro, A.; Mencarelli, F. Partial least squares discriminant analysis: A dimensionality reduction method to classify hyperspectral data. *Stat. Appl. Ital. J. Appl. Stat.* **2018**, *31*, 181–200. [CrossRef]
53. Krishnan, A.; Williams, L.J.; McIntosh, A.R.; Abdi, H. Partial Least Squares (PLS) methods for neuroimaging: A tutorial and review. *Neuroimage* **2011**, *56*, 455–475. [CrossRef]
54. Ray, S.S.; Das, G.; Singh, J.P.; Panigrahy, S. Evaluation of hyperspectral indices for LAI estimation and discrimination of potato crop under different irrigation treatments. *Int. J. Remote Sens.* **2006**, *27*, 5373–5387. [CrossRef]
55. Mehmood, T.; Sæbø, S.; Liland, K.H. Comparison of variable selection methods in partial least squares regression. *J. Chemom.* **2020**, *34*. [CrossRef]
56. Abdi, H. Partial least squares regression and projection on latent structure regression (PLS Regression). *WIREs Comput. Stat.* **2010**, *2*, 97–106. [CrossRef]
57. Lu, B.; He, Y. Evaluating empirical regression, machine learning, and radiative transfer modelling for estimating vegetation chlorophyll content using bi-seasonal hyperspectral images. *Remote Sens.* **2019**, *11*, 1979. [CrossRef]

58. Thenkabail, P.S.; Mariotto, I.; Gumma, M.K.; Middleton, E.M.; Landis, D.R.; Huemmrich, K.F. Selection of hyperspectral narrowbands (hnbs) and composition of hyperspectral twoband vegetation indices (HVIS) for biophysical characterization and discrimination of crop types using field reflectance and hyperion/EO-1 data. *IEEE J. Sel. Top. Appl. Earth Obs. Remote Sens.* **2013**, *6*, 427–439. [CrossRef]
59. Thenkabail, P.S.; Smith, R.B.; De Pauw, E. Hyperspectral Vegetation Indices and Their Relationships with Agricultural Crop Characteristics. *Remote Sens. Environ.* **2000**, *71*, 158–182. [CrossRef]
60. Thenkabail, P.; Lyon, J. (Eds.) *Hyperspectral Remote Sensing of Vegetation*; CRC Press: Boca Raton, FL, USA, 2012.
61. Durif, G.; Modolo, L.; Michaelsson, J.; Mold, J.E.; Lambert-Lacroix, S.; Picard, F. High dimensional classification with combined adaptive sparse PLS and logistic regression. *Bioinformatics* **2018**, *34*, 485–493. [CrossRef]
62. Deb, D.; Mackey, D.; Opiyo, S.O.; McDowell, J.M. Application of alignment-free bioinformatics methods to identify an oomycete protein with structural and functional similarity to the bacterial AvrE effector protein. *PLoS ONE* **2018**, *13*, 1–19. [CrossRef]
63. Sampaio, P.S.; Soares, A.; Castanho, A.; Almeida, A.S.; Oliveira, J.; Brites, C. Optimization of rice amylose determination by NIR-spectroscopy using PLS chemometrics algorithms. *Food Chem.* **2018**, *242*, 196–204. [CrossRef]
64. Li, J.; Tong, Y.; Guan, L.; Wu, S.; Li, D. Optimization of COD determination by UV–vis spectroscopy using PLS chemometrics algorithms. *Optik* **2018**, *174*, 591–599. [CrossRef]
65. Luedeling, E.; Gassner, A. Partial Least Squares Regression for analyzing walnut phenology in California. *Agric. For. Meteorol.* **2012**, *158–159*, 43–52. [CrossRef]
66. Song, S.; Gong, W.; Zhu, B.; Huang, X. Wavelength selection and spectral discrimination for paddy rice, with laboratory measurements of hyperspectral leaf reflectance Shalei. *ISPRS J. Photogramm. Remote Sens.* **2011**, *66*, 672–682. [CrossRef]
67. Mehmood, T.; Liland, K.H.; Snipen, L.; Sæbø, S. A review of variable selection methods in Partial Least Squares Regression. *Chemom. Intell. Lab. Syst.* **2012**, *118*, 62–69. [CrossRef]
68. Wang, Y.; Gao, Y.; Yu, X.; Wang, Y.; Deng, S.; Gao, J.-M. Rapid Determination of Lycium Barbarum Polysaccharide with Effective Wavelength Selection Using Near-Infrared Diffuse Reflectance Spectroscopy. *Food Anal. Methods* **2016**, *9*, 131–138. [CrossRef]
69. Chen, H.; Xia, Z. Determination of total flavonoids in propolis based on NIR spectroscopy technology. *Chin. J. Pharm. Anal.* **2014**, *34*, 1868–1873.
70. Eriksson, I.; Johansson, E.; Kettaneh-Wold, N.; Wold, S. *Multi- and Megavariate Data Analysis. Principles and Applications*; MKS Umetrics: Malmö, Sweden, 2002; Volume 16.
71. Ding, X.-B.; Zhang, C.; Liu, F.; Song, X.L.; Kong, W.W.; He, Y. Determination of soluble solid content in strawberry using hyperspectral imaging combined with feature extraction methods. *Spectroscopy Spectr. Anal.* **2015**, *35*, 1020–1024.
72. Pan, L.; Lu, R.; Zhu, Q.; Tu, K.; Cen, H. Predict compositions and mechanical properties of sugar beet using hyperspectral scattering. *Food Bioprocess Technol.* **2016**, *9*, 1177–1186. [CrossRef]
73. Guzmán, E.; Baeten, V.; Pierna, J.A.F.; García-Mesa, J.A. Application of low-resolution Raman spectroscopy for the analysis of oxidized olive oil. *Food Control.* **2011**, *22*, 2036–2040. [CrossRef]
74. Li, H.D.; Zeng, M.M.; Tan, B.-B.; Liang, Y.Z.; Xu, Q.S.; Cao, D.S. Recipe for revealing informative metabolites based on model population analysis. *Metabolomics* **2010**, *6*, 353–361. [CrossRef]
75. Pan, T.; Chen, W.; Chen, Z.; Xie, J. Wavelength selection for NIR spectroscopic analysis of chemical oxygen demand based on different partial least squares models. *Key Eng. Mater.* **2011**, *480–481*, 393–396. [CrossRef]
76. Zou, X.; Zhao, J.; Li, Y. Selection of the efficient wavelength regions in FT-NIR spectroscopy for determination of SSC of "Fuji" apple based on BiPLS and FiPLS models. *Vib. Spectrosc.* **2007**, *44*, 220–227. [CrossRef]
77. Yang, Q.; Zhu, G.; Ren, P.; Long, S.; Yang, J. Wavelength selection for NIR spectroscopic analysis of chemical oxygen demand based on different partial least squares models. *J. Anal. Sci.* **2016**, *32*, 485–489.
78. Fan, S.; Zhang, B.; Li, J.; Huang, W.; Wang, C. Effect of spectrum measurement position variation on the robustness of NIR spectroscopy models for soluble solids content of apple. *Biosyst. Eng.* **2016**, *143*, 9–19. [CrossRef]
79. Saeys, Y.; Inza, I.; Larrañaga, P. A review of feature selection techniques in bioinformatics. *Bioinformatics* **2007**, *23*, 2507–2517. [CrossRef]
80. Lê Cao, K.A.; Rossouw, D.; Robert-Granié, C.; Besse, P. A sparse PLS for variable selection when integrating omics data. *Stat. Appl. Genet. Mol. Biol.* **2008**, *7*. [CrossRef] [PubMed]
81. Bax, E.; Weng, L.; Tian, X. Speculate-correct error bounds for k-nearest neighbor classifiers. *Mach. Learn.* **2019**, *108*, 2087–2111. [CrossRef]

82. Grove, A.J. General Convergence Results for Linear Discriminant Updates. *Mach. Learn.* **2001**, *43*, 173–210. [CrossRef]
83. AbuZeina, D.; Al-Anzi, F.S. Employing fisher discriminant analysis for Arabic text classification. *Comput. Electr. Eng.* **2018**, *66*, 474–486. [CrossRef]
84. Pedregosa, F.; Varoquaux, G.; Gramfort, A.; Michel, V.; Thirion, B.; Grisel, O.; Blondel, M.; Prettenhofer, P.; Weiss, R.; Dubourg, V.; et al. Scikit-learn: Machine Learning in Python. *J. Mach. Learn. Res.* **2011**, *12*, 2825–2830.
85. Suarez, L.A.; Apan, A.; Werth, J. Detection of phenoxy herbicide dosage in cotton crops through the analysis of hyperspectral data. *Int. J. Remote Sens.* **2017**, *38*, 6528–6553. [CrossRef]
86. Zhou, Q.; Huang, W.; Fan, S.; Zhao, F.; Liang, D.; Tian, X. Non-destructive discrimination of the variety of sweet maize seeds based on hyperspectral image coupled with wavelength selection algorithm. *Infrared Phys. Technol.* **2020**, *109*, 103418. [CrossRef]
87. Xia, Z.; Zhang, C.; Weng, H.; Nie, P.; He, Y. Sensitive Wavelengths Selection in Identification of Ophiopogon japonicus Based on Near-Infrared Hyperspectral Imaging Technology. *Int. J. Anal. Chem.* **2017**, *2017*, 1–11. [CrossRef]
88. Snavely, N.; Seitz, S.M.; Szeliski, R. Modeling the world from Internet photo collections. *Int. J. Comput. Vis.* **2008**, *80*, 189–210. [CrossRef]
89. Mesas-Carrascosa, F.J.; de Castro, A.I.; Torres-Sánchez, J.; Triviño-Tarradas, P.; Jiménez-Brenes, F.M.; García-Ferrer, A.; López-Granados, F. Classification of 3D point clouds using color vegetation indices for precision viticulture and digitizing applications. *Remote Sens.* **2020**, *12*, 317. [CrossRef]
90. Mesas-carrascosa, F.J.; Rumbao, I.C.; Alberto, J.; Berrocal, B.; Porras, A.G. Positional Quality Assessment of Orthophotos Obtained from Sensors Onboard Multi-Rotor UAV Platforms. *Sensors* **2014**, *14*, 22394–22407. [CrossRef] [PubMed]
91. Barreto, M.A.P.; Johansen, K.; Angel, Y.; McCabe, M.F. Radiometric assessment of a UAV-based push-broom hyperspectral camera. *Sensors* **2019**, *19*, 4699. [CrossRef] [PubMed]
92. van Zyl, J.J. The Shuttle Radar Topography Mission (SRTM): A breakthrough in remote sensing of topography. *Acta Astronaut.* **2001**, *48*, 559–565. [CrossRef]
93. Van Rossum, G.; Drake, F.L. *The Python Language Reference*; Python Software Foundation: Scotts Valley, CA, USA, 2011; p. 109.
94. Degenhardt, F.; Seifert, S.; Szymczak, S. Evaluation of variable selection methods for random forests and omics data sets. *Brief. Bioinform.* **2019**, *20*, 492–503. [CrossRef]
95. Tibshirani, R. Regression Shrinkage and Selection via the Lasso. *J. R. Stat. Soc.* **1996**, *58*, 267–288. [CrossRef]
96. Jain, A. A Complete Tutorial on Ridge and Lasso Regression in Python. Available online: https://www.analyticsvidhya.com/blog/2016/01/complete-tutorial-ridge-lasso-regression-python/ (accessed on 5 August 2020).
97. Kursa, M.B.; Rudnicki, W.R. Feature selection with the boruta package. *J. Stat. Softw.* **2010**, *36*, 1–13. [CrossRef]
98. Alba-Fernández, M.V.; Ariza-López, F.J.; Rodríguez-Avi, J.; García-Balboa, J.L. Statistical methods for thematic-accuracy quality control based on an accurate reference sample. *Remote Sens.* **2020**, *12*, 816. [CrossRef]
99. R Core Team. *R: A Language and Environment for Statistical Computing*; R Core Team: Vienna, Austria, 2018.
100. RStudio Team. *RStudio: Integrated Development for R*; RStudio: Boston, MA, USA, 2017.
101. Mevik, B.H.; Wehrens, R. The pls package: Principal component and partial least squares regression in R. *J. Stat. Softw.* **2007**, *18*, 1–23. [CrossRef]
102. Northrop, P.J.; Attalides, N.; Jonathan, P. Cross-validatory extreme value threshold selection and uncertainty with application to ocean storm severity. *J. R. Stat. Soc. Ser. C Appl. Stat.* **2017**, *66*, 93–120. [CrossRef]
103. Kucheryavskiy, S. Mdatools—R package for chemometrics. *Chemom. Intell. Lab. Syst.* **2020**, *198*. [CrossRef]
104. Li, H.D.; Xu, Q.S.; Liang, Y.Z. LibPLS: An integrated library for partial least squares regression and linear discriminant analysis. *Chemom. Intell. Lab. Syst.* **2018**, *176*, 34–43. [CrossRef]
105. Chun, H.; Keleş, S. Sparse partial least squares regression for simultaneous dimension reduction and variable selection. *J. R. Stat. Soc. Ser. B Stat. Methodol.* **2010**, *72*, 3–25. [CrossRef]
106. Friedman, J.; Trevor, H.; Tibshirani, R. Regularization paths for generalized linear models via coordinate descent. *J. Stat. Softw.* **2010**, *33*, 1–22. [CrossRef]
107. Calderón, R.; Navas-Cortés, J.A.; Lucena, C.; Zarco-Tejada, P.J. High-resolution airborne hyperspectral and thermal imagery for early detection of Verticillium wilt of olive using fluorescence, temperature and narrow-band spectral indices. *Remote Sens. Environ.* **2013**, *139*, 231–245. [CrossRef]

108. Zarco-Tejada, P.J.; Camino, C.; Beck, P.S.A.; Calderon, R.; Hornero, A.; Hernández-Clemente, R.; Kattenborn, T.; Montes-Borrego, M.; Susca, L.; Morelli, M.; et al. Previsual symptoms of Xylella fastidiosa infection revealed in spectral plant-trait alterations. *Nat. Plants* **2018**, *4*, 432–439. [CrossRef]
109. Starzacher, A.; Rinner, B. Evaluating KNN, LDA and QDA classification for embedded online feature fusion. In Proceedings of the 2008 International Conference on Intelligent Sensors, Sensor Networks and Information Processing, Sydney, NSW, Australia, 15–18 December 2008; IEEE: Piscataway Township, NJ, USA, 2008; pp. 85–90.

Publisher's Note: MDPI stays neutral with regard to jurisdictional claims in published maps and institutional affiliations.

Article

Multi-Temporal Predictive Modelling of Sorghum Biomass Using UAV-Based Hyperspectral and LiDAR Data

Ali Masjedi [1,*], Melba M. Crawford [1,2], Neal R. Carpenter [3] and Mitchell R. Tuinstra [2]

1 Lyles School of Civil Engineering, Purdue University, West Lafayette, IN 47907, USA; mcrawford@purdue.edu
2 Department of Agronomy, Purdue University, West Lafayette, IN 47907, USA; mtuinstr@purdue.edu
3 Bayer US-Crop Science, Chesterfield, MO 63017, USA; neal.carpenter@bayer.com
* Correspondence: amasjedi@purdue.edu; Tel.: +1-765-772-8309

Received: 13 September 2020; Accepted: 24 October 2020; Published: 1 November 2020

Abstract: High-throughput phenotyping using high spatial, spectral, and temporal resolution remote sensing (RS) data has become a critical part of the plant breeding chain focused on reducing the time and cost of the selection process for the "best" genotypes with respect to the trait(s) of interest. In this paper, the potential of accurate and reliable sorghum biomass prediction using visible and near infrared (VNIR) and short-wave infrared (SWIR) hyperspectral data as well as light detection and ranging (LiDAR) data acquired by sensors mounted on UAV platforms is investigated. Predictive models are developed using classical regression-based machine learning methods for nine experiments conducted during the 2017 and 2018 growing seasons at the Agronomy Center for Research and Education (ACRE) at Purdue University, Indiana, USA. The impact of the regression method, data source, timing of RS and field-based biomass reference data acquisition, and the number of samples on the prediction results are investigated. R^2 values for end-of-season biomass ranged from 0.64 to 0.89 for different experiments when features from all the data sources were included. Geometry-based features derived from the LiDAR point cloud to characterize plant structure and chemistry-based features extracted from hyperspectral data provided the most accurate predictions. Evaluation of the impact of the time of data acquisition during the growing season on the prediction results indicated that although the most accurate and reliable predictions of final biomass were achieved using remotely sensed data from mid-season to end-of-season, predictions in mid-season provided adequate results to differentiate between promising varieties for selection. The analysis of variance (ANOVA) of the accuracies of the predictive models showed that both the data source and regression method are important factors for a reliable prediction; however, the data source was more important with 69% significance, versus 28% significance for the regression method.

Keywords: high-throughput phenotyping; hyperspectral data; LiDAR; biomass prediction

---

1. Introduction

Biomass yield is an important trait of biofuel crops such as sorghum, as it is a key factor in determining the amount of biofuel that can be produced. With recent advances in science and technology surrounding genotyping, it has become possible to create numerous genotypes of a plant [1] and then select the genotypes with the maximum biomass production. However, traditional methods of biomass measurement involving labor-intensive and time-consuming destructive sampling do not meet the requirements for timely evaluation of the genotypes in large-scale breeding programs. Recently, remote sensing (RS) data have been explored for estimation of many phenotypic traits, including leaf area index (LAI) [2,3], canopy height [4,5], nitrogen content [6], and biomass [7–11], to replace traditional in-field phenotyping.

Sensors on satellites and manned aircraft can provide data with high spectral resolution, but the spatial and temporal resolutions are inadequate for agricultural breeding programs that are based on small plots. Remote sensing via unmanned aerial vehicles (UAV) is currently being investigated as a means to close the gap because of its capability to acquire the high temporal and spatial resolution data required for high throughput phenotyping over relatively limited areas. UAVs can collect huge quantities of data "on demand", providing opportunities for estimation and prediction of a wide range of agronomic traits [12–24].

In this study, remotely sensed biomass prediction of varieties of sorghum is investigated using the data acquired by RGB, hyperspectral and LiDAR sensors mounted on UAV platforms. Sorghum has attracted attention in recent years, both for its broad-based potential usage and its drought and heat tolerance. The grain of some varieties is now used for human consumption and animal feed in developed, as well as developing countries. Recently, some varieties of sorghum have been developed as an energy crop that can produce reasonable quantities of ethanol [25]. Sorghum has an annual growth cycle, high calorific value, and low management cost [25], making it an efficient biofuel crop. Many studies focus on developing enhanced genotypes that can produce more energy-rich plant material (biomass) [26]. It is important for these breeding studies to predict the end-of-season yield biomass of the planted varieties as soon as possible in the growing season to screen varieties, and thus reduce investment of expensive resources in monitoring for the whole season.

Biomass prediction based on data analytics models and RS data is challenging for multiple reasons, including (1) the complex relationship between biomass and RS data [27], (2) limited number of ground reference samples for developing and validating models for an experiment [28], and (3) high variability between the samples in an experiment [29]. Moreover, the relationships between the RS-based features and traits vary across the growing season [27]. Extraction of robust, explanatory features as predictors of the trait of interest is critical to development of machine learning models. A small number of field reference samples relative to the number of features (and thereby potentially the number of parameters to estimate) is a difficult issue for remote sensing-based phenotyping. Unfortunately, reference sampling data is time-consuming and expensive to collect in agricultural fields.

In this study, the objective is to develop baseline predictive models for sorghum biomass yield based on classical machine learning methods using multi-date remote sensing and ground reference data. The impact of timing of the data acquisition relative to days since sowing and the importance of the features extracted from the data are also investigated.

The remainder of this paper is organized as follows. Section 2 surveys the literature on the predictive models based on remote sensing data. In Section 3, the study area, reference data, and remote sensing data are described. Additionally, the methodologies including feature extraction, regression models, as well as statistical analysis are explained. Experimental results are presented in Section 4 and discussed in Section 5, and finally, conclusions are drawn in Section 6.

2. Related Work

Many studies have explored the potential for prediction or estimation of phenotypic traits utilizing spectral data. Potgieter et al. [2] found that indices obtained from a UAV-based multispectral sensor over a sorghum field with two different genotypes were correlated with the LAI measured in the field. Using spectrometer measurements acquired in wheat fields in different locations at multiple times during three growing seasons, Feng et al. [30] demonstrated that the nitrogen content of the leaves are highly correlated with the parameters derived from the first derivative of reflectance. The authors investigated correlation between phenotypic traits and remote sensing-based features. Estimation of the values of quantitative traits such as biomass at a given date, or prediction at a future date based on earlier data are more difficult.

Researchers have developed predictive models based on various remote sensing inputs and modeling approaches. Foster et al. [31] compared the performance of partial least squares regression (PLSR) and linear regression models in estimating biomass of a high-biomass sorghum variety. They concluded that PLSR can provide more accurate predictions using the normalized difference vegetation index (NDVI) calculated from field spectrometer measurements collected in July (three months after sowing). Yue et al. [32] also demonstrated that the PLSR provided the best results among the eight regression techniques investigated for wheat biomass estimation. Fassnacht et al. [33] investigated the importance of the prediction method, as well as sample size and sensor type for biomass predictions in forest environments, with the best results being obtained with a random forest (RF) model. Using airborne LiDAR and spaceborne hyperspectral data, the authors concluded that for their experiments, the sensor type was the most important factor in the prediction accuracy.

Multiple studies have also investigated biomass prediction using LiDAR data [8,34–43]. Harkel et al. [35] evaluated the accuracy of biomass prediction using LiDAR data for various crops. In [41], the use of LiDAR combined with spectral vegetation indices (VI) derived from multispectral data provided more accurate biomass estimates than LiDAR and multispectral data individually. Luo et al. [42] extracted various features, including variables from discrete-return LiDAR, LiDAR pseudo-waveform, and VIs from hyperspectral imagery and used them to predict biomass in a RF model. They showed that the combined data have potential for improving predictions of crop parameters. Other studies have shown that fusion of airborne-based hyperspectral and LiDAR data provided better results than those achieved using data from either individual sensor type [44].

Most studies have developed predictive models for a limited number of experiments, each including only a few genotypes of a crop, although in breeding programs, hundreds or thousands of genotypes with high variability in biomass, as well as spectral and structural characteristics, are included in each experiment. One contribution of this study is that extensive data are acquired consistently over large breeding trials. Predictive models are developed for nine distinct experimental trials conducted over two years and include thousands of genotypes of sorghum. In this study, the objective is to provide a robust framework for predicting sorghum biomass which is suitable for plant breeding research and industrial applications. To accomplish this, we: (1) evaluate the importance of the features extracted from multiple data sources; (2) evaluate multiple prediction models; (3) investigate the impact of various sorghum genotypes on prediction accuracy; (4) investigate the model performance for early, mid, and late season biomass prediction; (5) investigate the impact of the timing of the RS data acquisition on prediction relative to days since sowing; and (6) evaluate the impact of the number of training samples on the prediction accuracy.

3. Materials and Methods

3.1. Experimental Site

The field experiments were conducted over two years in approximately 2.8 ha sorghum breeding trials in different fields at the Purdue University Agronomy Center for Research and Education (ACRE) farm (see Figure 1). There were four distinct trials in 2017: the hybrid calibration (HyCal-17), the inbred calibration (InCal-17), the sorghum biodiversity (SbDiv-17), and the sorghum bioenergy (SbBAP-17) panels. In 2018, the field experiments consisted of five distinct trials: the hybrid calibration (HyCal-18), the inbred calibration (InCal-18), the inbred calibration test cross (InCalTc-18), the sorghum biodiversity test cross (SbDivTc-18), and the sorghum nitrogen test (SNitTs-18). The experiments were conducted using randomized complete block designs and planted at 220,000 plants per ha. The commercial hybrid varieties planted in the HyCal-17 and HyCal-18 experiments are listed in Table A1 in Appendix A. The RGB images of the field trials are shown in Figures A1 and A2 in Appendix B.

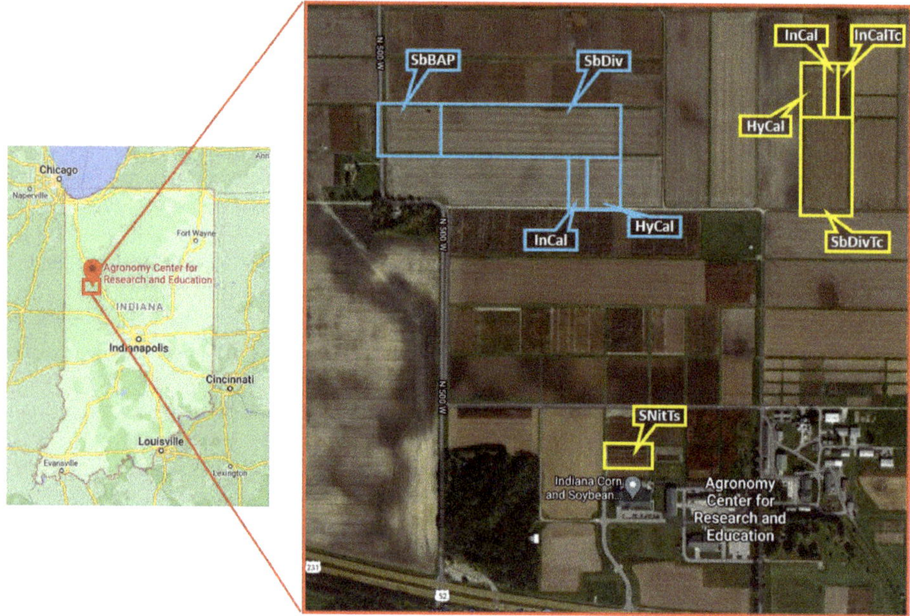

Figure 1. Experimental site: the field trials at Agronomy Center for Research and Education (ACRE) are highlighted by blue (the 2017 fields) and yellow (the 2018 fields).

3.2. Ground Reference Data

Details of the experiment trials including sowing date, which differed by approximately one week, and harvest dates are provided in Table 1. For the HyCal-17 and HyCal-18 panels, biomass data were destructively collected multiple times during each growing season. For all other experiments, the biomass data were collected only once at the end of each growing season using a two-row combine harvester. The weight of the shredded plant material of each plot was considered as the fresh biomass value for that plot. After harvesting, around 500 g of the shredded plant material was used to determine the moisture content of each plot by measuring the fresh weight and dry weight (after drying the plant materials).

Table 1. Experiment designs for the 2017 and 2018 growing seasons.

Trial	Year	Genotype	# of Plots	# of Genotypes	Sowing Date	Harvest Date	Available Biomass Data
HyCal-17	2017	Hybrid	72	18	16/05	27/09	27/06, 17/07, 31/07, 08/08, 27/09
InCal-17	2017	Inbred	120	60	16/05	27/09	27/09
SbBAP-17	2017	Inbred	760	350	16/05	28/09	28/09
SbDiv-17	2017	Inbred	1800	840	17/05	09/11	09/11
HyCal-18	2018	Hybrid	72	18	08/05	09/08	27/06, 12/07, 09/08
InCal-18	2018	Inbred	108	54	08/05	09/08	09/08
InCalTc-18	2018	Hybrid	108	54	08/05	06/08	06/08
SbDivTc-18	2018	Hybrid	1260	630	08/05	02/08	02/08 and 14/08
SNitTs-18	2018	Inbred	112	4	04/06	02/10	02/10

Figure 2 shows the distribution of the fresh biomass values for the experimental trials in the 2017 and 2018 growing seasons. Figure 2a,b shows the fresh biomass distribution of the HyCal-17 and

HyCal-18 panels, respectively, during the growing seasons. In 2017, the biomass data on 27 June, 17 July, and 7 August were collected by hand harvesting one meter sections of three rows from each plot. Harvesting for all other dates was performed with a two-row combine harvester. These figures indicate that the genotypes have similar biomass early in the season but differ at the end of the season. Figure 2c shows the distribution of the end of the season biomass of the nine trials over both years: (1) the InCal-17 and SbDiv-17 are similar, (2) the HyCal-17 and HyCal-18 have similar shapes but different ranges of values, and (3) SbDivTc-18 and HyCal-17 are similar in both shape and range of values.

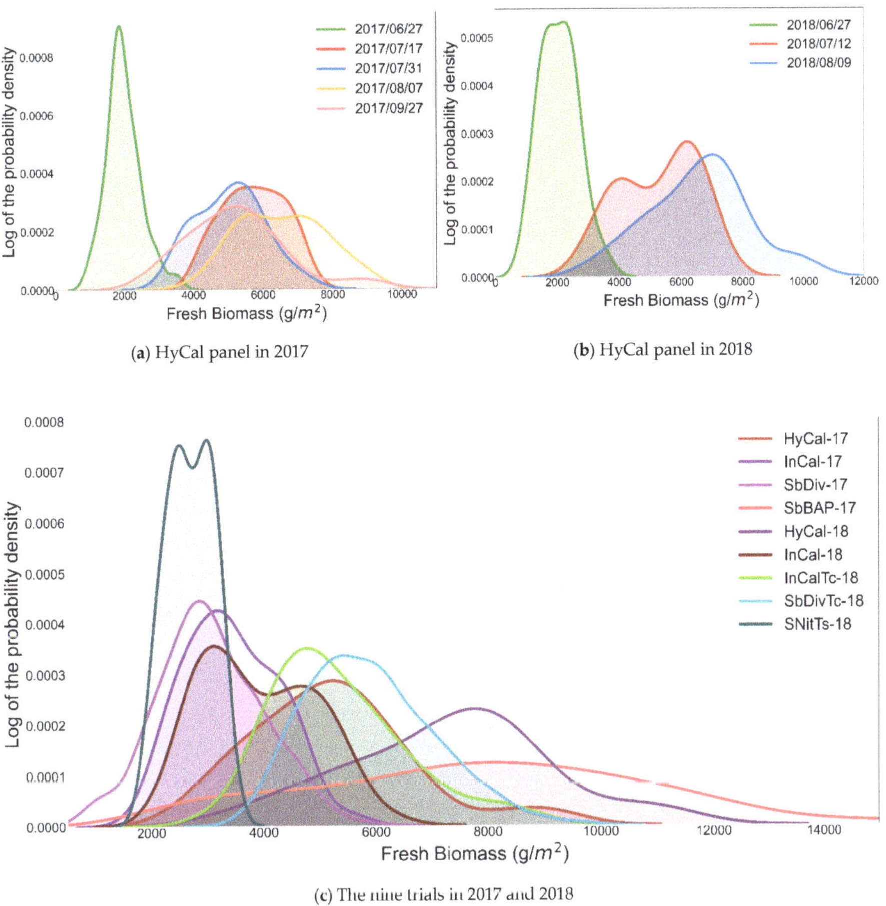

(**a**) HyCal panel in 2017

(**b**) HyCal panel in 2018

(**c**) The nine trials in 2017 and 2018

Figure 2. Distribution of the fresh biomass data in the nine trials in the 2017 and 2018 growing seasons.

3.3. Remote Sensing Data

This study includes RGB, hyperspectral, and LiDAR remote sensing data collected by custom designed UAV platforms. All remote sensing data acquisition platforms were flown with global navigation satellite system/inertial navigation system (GNSS/INS) units for direct georeferencing. The description of the sensors used in this study is provided in Table 2. RGB data for this study were collected using a Sony Alpha ILCE-7R RGB camera delivering high-resolution UAV-based aerial imagery. LiDAR data were collected with a Velodyne VLP-16 3D LiDAR sensor operating in the strongest return mode providing an average point cloud density of 750 points per m^2. Both the RGB camera and the VLP-16 sensor are mounted on a DJI Matrice 600 Pro (M600P) platform. Spatial and temporal system calibration for the datasets used in this study were conducted using the approaches described in [45] and [46], respectively. Additionally, the georeferenced orthomosaics were generated using the structure from motion strategies introduced in [47,48].

Table 2. Sensor Descriptions.

Sensor	Description
RGB	Sony Alpha ILCE-7R Sony 35mm Lens Full-frame 36.4MP
LiDAR	Velodyne VLP-16 600 rotations per minute (RPM), 360-degree horizontal FOV Maximum range of 100 m
VNIR	Headwall Photonics Nano-Hyperspec imaging sensor 272 spectral bands at 2.2 nm/band from 400 nm to 1000 nm 640 spatial channels at 7.4 µm/pixel, 12 mm lens (in 2017) and 8 mm lens (in 2018)
SWIR	Headwall Photonics Micro-Hyperspec pushbroom 166 spectral bands at 10 nm/band from 900 nm to 2500 nm 384 spatial channels at 24 µm/pixel, 25 mm lens

Visible near infrared (VNIR) and short wave infrared (SWIR) hyperspectral data were collected with two Headwall Photonics push-broom scanners. In 2017, the VNIR sensor was flown at an altitude of 60 m with a 12 mm Schneider lens, resulting in a ground sampling distance (GSD) of ~4 cm. An 8 mm lens was used in 2018, and the flying height was 40 m to maintain the GSD at ~4 cm and accommodate the field of view of other sensors on the platform. In both years, the SWIR sensor was flown with a 25 mm lens at 40 m, resulting in approximately a 4 cm GSD. In 2018, the VNIR and SWIR sensors were integrated and flown together on a single UAV platform. A rigorous boresight calibration process described by Habib et al. [49] was applied, yielding simultaneously collected co-aligned VNIR and SWIR data. Similar to [50], all the hyperspectral data were converted to reflectance using the empirical line method to relate the spectra collected from the UAV to data acquired by an SVC 1024i field spectrometer over the calibration targets placed in the field for each acquisition. The data acquired by the sensors in both 2017 and 2018 are listed in Table A2 in Appendix A.

3.4. Feature Extraction

As discussed earlier, it is important to extract features that are related to the specific trait of interest and are preferably not redundant. In this study, both traditional and new candidate features focused on the relevance to biomass prediction were extracted from rows 2 and 3 of the 4- or 12-row plots to minimize the border effect. For each acquired data set (listed in Table A2 in Appendix A), we extracted the features described in the following sections.

3.4.1. Hyperspectral-Based Features

- Spectral Reflectance

From the Hyperspectral Imaging (HSI) data, the average reflectance values of the plots were calculated from rows 2 and 3 of each plot after masking the shadow and soil pixels.

- Vegetation Indices

Vegetation Indices (VIs) obtained from HSI data have been widely used in different applications, as they are computationally simple and representative of the relevant chemically interpretable absorption and reflectance features in the spectrum. In this study, 13 vegetation indices, listed in Table A3 in Appendix A were extracted and used in the predictive models.

- Integration Features

NIR bands are particularly important for representing plant physiology but are subject to the time during the growing season and environmental conditions. The area under the spectral curve for a given range from λ_a to λ_b is defined as $Intg(\lambda_a, \lambda_b) = \int_{\lambda_a}^{\lambda_b} S(\lambda)d\lambda$, where $S(\lambda)$ is the reflectance at λ nm. Using different ranges of spectral values, six features were extracted from each HSI spectrum as listed in Table A4 in Appendix A.

- Derivative Features

The spectral derivatives, which quantify slope, curvature, and higher-order aspects of reflectance spectra, can be useful by revealing spectral features that may not be apparent in reflectance data alone [51]. For example, the "red-edge" position (between 680 nm and 750 nm) in crop reflectance data can be easily identified in the derivative spectra, and has been related to crop biomass [52]. Feng et al. analyzed 20 spectral derivative features near the red edge area to estimate wheat leaf nitrogen concentration [53]. In this study, the polished spectra were calculated using a Savitzky–Golay filter [54], then the first derivative (FDR) and second derivative (SDR) of the spectra were extracted. From FDR and SDR, 11 features were extracted and used in the additional analysis as described in Table A5 in Appendix A. These features were selected at wavelengths where spectra of the varieties differed and were also uncorrelated.

3.4.2. LiDAR-Based Features

The 3D structural characteristics of the plants in a plot can be described using various features extracted from LiDAR data. The digital terrain model (DTM) was derived from LiDAR data acquired before the emergence of the plants in each field by interpolating the LiDAR point cloud into a regular grid (8 × 8 cm in this study) using the nearest neighbour interpolation method. The DTM represents the bare earth height information and is assumed to be constant throughout the growing season. For each point cloud data acquired throughout the season, the height of points in the was estimated by subtracting the DTM from the "z" coordinate of each point. Then, the following features were extracted from the point cloud of each plot:

- Height Percentile

To capture the vertical distribution of the LiDAR points in each plot, the 30th, 50th, 75th, 90th, and 95th percentile height values from the point cloud of each plot were calculated.

- Canopy Volume

To estimate volume related characteristics of the canopy in each plot, a grid with cells of size 8 × 8 cm was assigned to each plot, and then the associated height was calculated from the points located in each cell, multiplied by the size of the cell to estimate the volume of the canopy within each cell. The aggregate "volume" in each plot is referred to as the volume of the vegetation within a plot. The height of each cell in this study was calculated as the average of the height of the lowest point and the height of the highest point in each cell.

- Canopy Cover

Canopy cover can be estimated from LiDAR data as the ratio of above-ground points (or, canopy points) to the total number of LiDAR points in a given area. The following approach was used in this study for canopy cover estimation for each plot. First, the field is divided into grid cells of a user-defined dimensions (8 × 8 cm in this study, consistent with the canopy volume calculation). Then, for each grid cell, the LiDAR points are split into two groups, canopy points and bare earth based on their height using a user-defined threshold. The points above the threshold are considered as canopy points. The canopy cover is estimated as the ratio of the number of canopy points to the total number of LiDAR points in each cell. The average of the canopy cover estimated for the cells located in each plot is assigned as the canopy cover for that plot. In this study, candidate threshold values were 0.1, 0.2, 0.3, 0.4, 0.5, and 0.75 multiplied by the 95th percentile height of each plot, resulting in six height-dependent canopy cover related features.

- Height Statistics

The spatial distribution of the height of the LiDAR points in each plot can also be represented using statistical moments of the distribution. These statistics can also be included as candidate input features for predictive models.

3.5. Regression-Based Modeling Approaches

Common regression-based approaches such as partial least squares regression (PLSR), support vector regression (SVR), and random forests (RF) are widely utilized to build predictive models with remote sensing based inputs. PLSR reduces a potentially large number of measured collinear input variables to a few uncorrelated latent variables while seeking to explain the maximum multi-dimensional variance of the dependent variable via a linear model. PLSR has been investigated for developing predictive models to estimate leaf biochemical and biophysical properties [55], chlorophyll content [56], carotenoid content [57], relative water content [58], protein, lignin, and cellulose [59], leaf nitrogen content [60], LAI [61], and biomass [34].

SVR is a supervised non-parametric regression technique, and therefore, no assumptions regarding the underlying data model are required. The original feature space is transformed into a higher dimensional space [62], with the goal of finding a hyperplane to predict the training data set. The optimal values of the kernel function parameters were obtained in this study by a general k-fold cross-validation in a grid search.

Random Forest (RF) modeling is an ensemble learning technique which uses a large set of classification and regression trees (CART) to predict the variable of interest [63]. In random forest regression, each tree is built by randomly choosing a set of variables and a subset of training samples with replacement. The selected samples are used for training, and the remaining observations are used in an internal cross-validation process to determine the performance of the RF model. The outputs of all trees are aggregated to produce a final prediction. A review of RF modeling in remote sensing applications is available in [64]. Similar to SVR, the parameter optimization was accomplished by a general k-fold cross-validation in a grid search method. A grid search method was used to select the best hyperparameters for each model. Table A6 lists the candidate parameters that were tested for each method in the grid search process. The Anaconda Distribution of Python version 3.7 with the Scikit-learn library [65] was used for conducting grid search and developing regression models.

3.6. Statistical Analysis

The one-way analysis of variance (ANOVA) is used to determine whether there is a significant difference among the groups of data. If there is a significant difference, then an honest significant difference (HSD) Tukey test (with $\alpha = 0.05$) can be conducted to determine which groups are significantly different from each other. We also use two-way ANOVA, to evaluate combinations of several variables or factors to identify those that have a significant effect on the estimates [66]. Prior to these statistical tests, the normality and homoscedasticity were confirmed by visually inspecting the variables. The statsmodels library [67] was used for data preparation and statistical analysis.

4. Results

4.1. Data Screening

4.1.1. Time Series of Biomass Data

In the 2017 and 2018 growing seasons, the destructive biomass data were collected multiple times (approximately every month) from the hybrid calibration panels. Figure 3 shows the fresh weight and moisture content of the 18 sorghum varieties planted in the HyCal-17 and HyCal-18 experiments. The moisture content increases at the beginning of the season until it reaches its maximum at 50 to 60 days after sowing. The fresh weight of the plants also increases rapidly at the beginning of the season, while at the end of the season, it decreases as the plants senesce. Among the varieties shown in Figure 3, those that are photoperiod sensitive ("Sordan Headless" and "Trudan Headless") did not flower in the environment in which these experiments were conducted, and continued to add plant material until the end of the season.

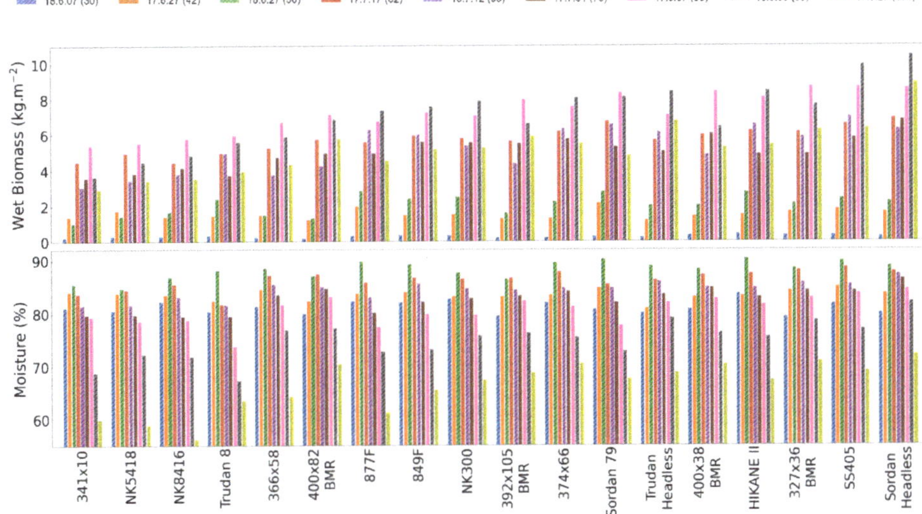

Figure 3. Ground reference data collected during the 2017 and 2018 growing seasons in the HyCal panels. For each variety, the data samples are sorted based on the day after sowing (DAS).

Each variety has four replicates in the hybrid Calibration panels, providing adequate sample data to compare the relationship across the varieties and associated changes during the growing season. For each date that the biomass data were collected, an ANOVA test was conducted on the fresh biomass, and if it indicated variability among the varieties was highly significant, a Tukey's multi-comparison test was performed. Figure 4 shows the results of Tukey's pairwise multi-comparison test ($\alpha = 0.05$) for the fresh biomass data collected in the 2017 and 2018 growing seasons. In general, at the beginning of each season, only a few varieties were significantly different, while the variability among the varieties at the end of each growing season was greater. From Figure 4 it is also clear that the two photoperiod sensitive varieties (varieties 16 and 18) were significantly different from the other varieties at the end of both 2017 and 2018 growing seasons.

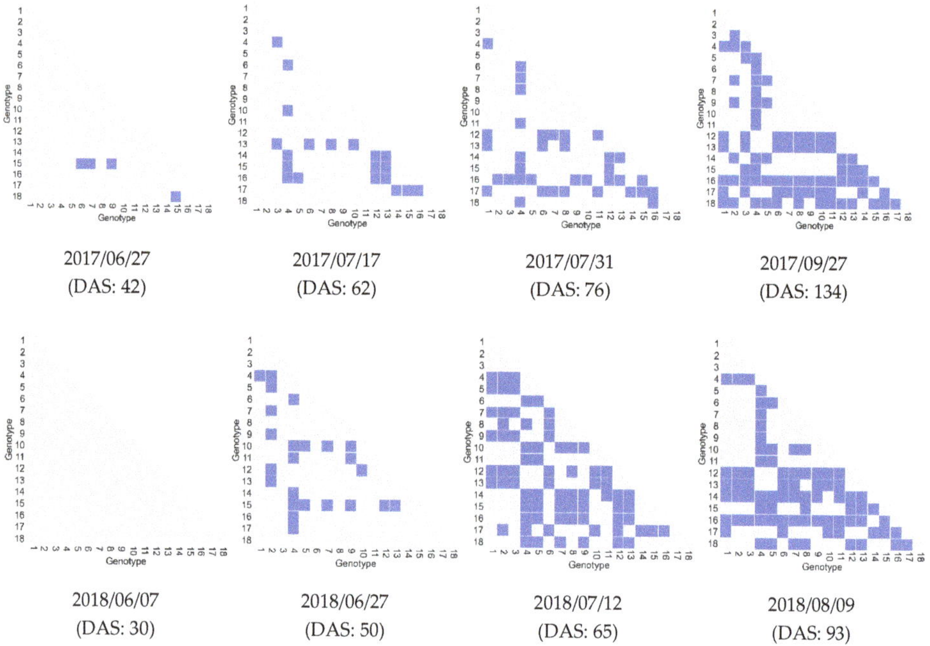

Figure 4. Tukey's pairwise multi-comparison test for the fresh biomass data collected in the 2017 and 2018 growing seasons for the 18 genotypes planted in HyCal-17 and HyCal-18 experiments. Blue indicates that the two varieties are significantly different ($\alpha = 0.05$).

4.1.2. Time Series of Remote Sensing Data

RS spectral signatures varied both across phenotypes and during the growing season. Figure 5 shows the average spectra of the 18 varieties of sorghum in the HyCal-18 panel on 18 July 2018 (day after sowing (DAS) = 71) (see Figure A4 in Appendix B for the variance plot). On that day, the signatures of the 18 varieties were very similar in the visible range of the spectrum, but there was more variability in the NIR portion of spectrum that may reflect variation in biochemical features, including lignin type and composition associated with the brown-midrib (bmr) traits. Figure 6 shows the reflectance of one of the varieties from June to September which shows there is very little change in the visible range of the spectrum, and especially in the blue and green bands, while the reflectance in NIR bands changes from about 35% to 60% on average (see Figure A5 in Appendix B for the variance plot). The maximum reflectance values were observed in the range of 750–850 nm on 3 July (DAS = 56). One of the reasons for changes in the reflectance for sorghum is the appearance of the panicles, which emerge a few days before the flowering date (the date on which 50% of panicles in a plot are flowered). Field notes indicate that the flowering date for "Trudan 8" was 10 July 2018.

Figure 7 shows the Digital Surface Model (DSM) generated from the LiDAR point cloud for the SbDivTc-18 panel from multiple dates in the 2018 growing season. From Figure 7, the plots are more similar at the beginning of the season, with greater differences later in the season. Figure 8 shows the point cloud data for two plots (rows 2 and 3) of the HyCal-18 experiment for multiple dates in the 2018 growing season. "341 × 10" is a dwarf grain sorghum variety (Figure 3), while "Trudan Headless" is a photoperiod sensitive forage sorghum with high biomass accumulation.

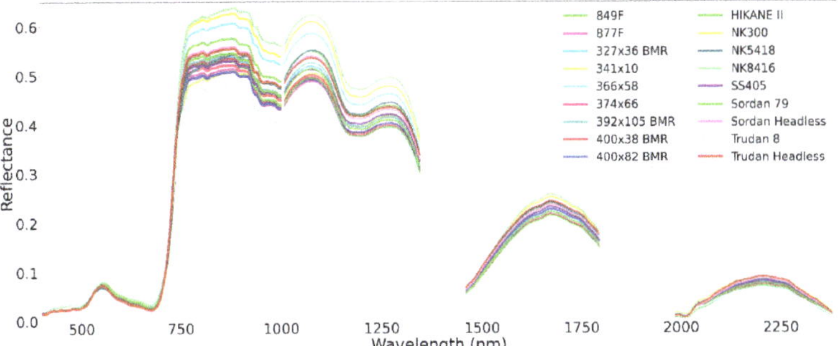

Figure 5. Spectra of the 18 Sorghum Varieties in HyCal-18 panel on 18 July 2018. The varieties are very similar in the visible range of the spectrum, but substantial variability is observed in the near infrared (NIR) portion of spectrum.

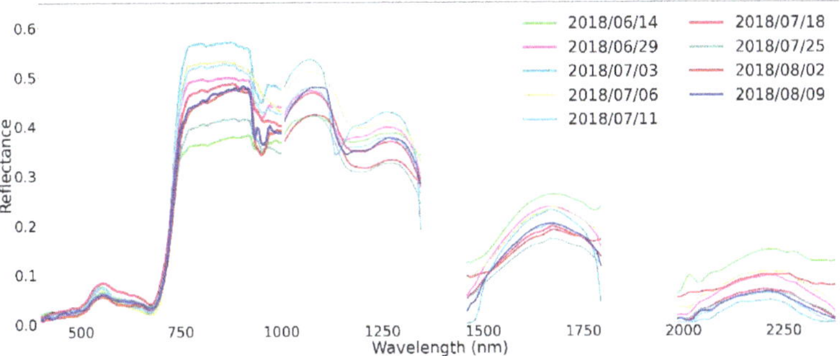

Figure 6. Example of reflectance of one of the varieties in HyCal-18 experiment ("Trudan 8") during the 2018 growing season.

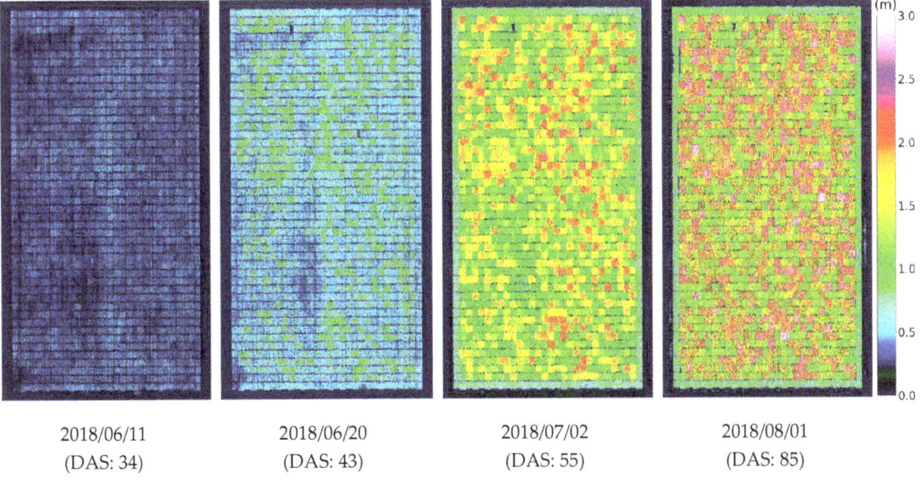

Figure 7. The 8 cm resolution DSM for multiple dates in the 2018 growing season for the SbDivTc experiment.

Figure 8. Point cloud data for rows two and three of two plots of the HyCal-18 experiment for multiple dates in the 2018 growing season. "341 × 10" and "Trudan Headless" achieve maximum height of 1.4 m and 3.2 m, respectively.

Figure 9 shows the mean and standard deviation of the height of all the plots for all the experiments in the 2017 and 2018 growing seasons extracted from the LiDAR point clouds, providing structural characteristics of the varieties planted in each experiment. For both HyCal-17 and HyCal-18, the height increases as the headless varieties continue to grow until the end of the season. The SbBAP-17 experiment also has the maximum average height at the end of the season, as it includes many plots of photoperiod sensitive genotypes that do not flower. The InCal-17, InCal-18, and SbDiv-17 include similar inbred varieties; thus, they have a very similar average height (also the lowest height values among the experiments). As was noted earlier, a histogram of the height of the points from the LiDAR point cloud provides information about the distribution of different height values in a plot. This genotype dependent information may be discriminating in predictive models. Figure A3 in Appendix B shows the histograms for the dwarf grain sorghum 341 × 10 and the photoperiod sensitive Trudan Headless varieties in the HyCal-17 experiment.

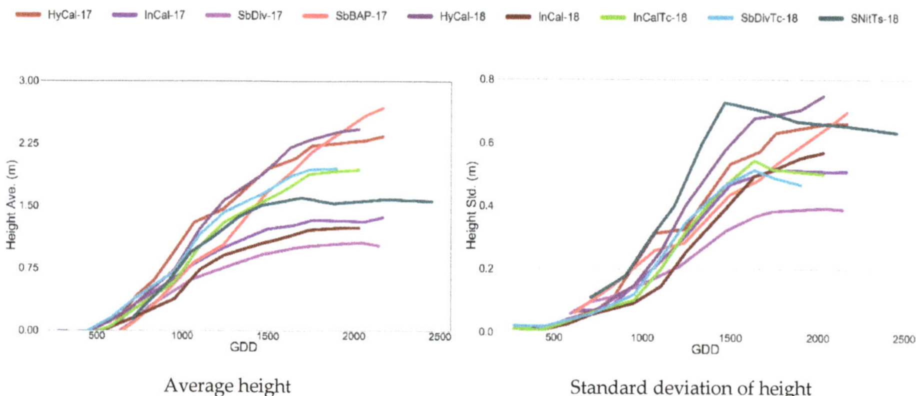

Figure 9. Average and standard deviation of the height of all the plots in each experimental trial in the 2017 and 2018 growing seasons.

The multi-temporal and cross-correlations during the growing season can be useful for screening for redundant features. Figure 10 shows the correlation matrices for the OSAVI and LiDAR-based canopy cover calculated using the combined data acquired over the HyCal-17 experiment in the 2017 growing season. It illustrates the rapid changes at the beginning of the season, especially prior to the flowering time (second week of July for this experiment) which is associated with the rapid growth of the plants. From Figure 10, OSAVI changed more than the canopy cover during the early season, and end of season OSAVI values have lower inter-temporal correlation compared to the correlation between the canopy cover values on corresponding dates.

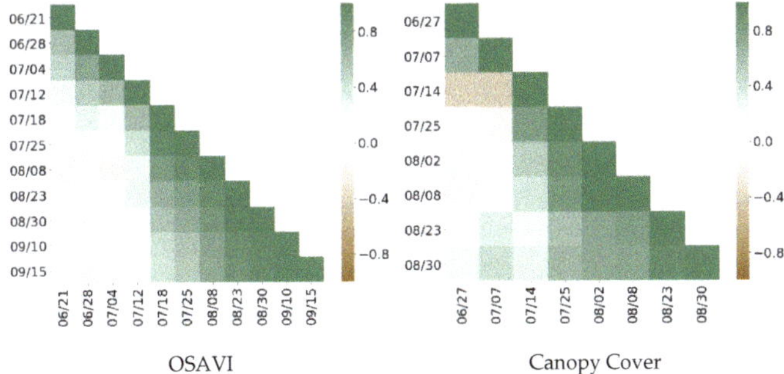

Figure 10. Correlation matrix calculated using all the remotely sensed hyperspectral data over the HyCal-17 experiment on different dates in the 2017 growing season. Note that as the light detection and ranging (LiDAR) and hyperspectral sensors were flown on separate platforms, the number of available data sets differs, and the data were not always collected on the same day.

Similar to the last section, Figures 11 and 12 show the results of the multiple-comparison Tukey's test for the OSAVI and volume features for the HyCal-17 and HyCal-18 experiments on multiple dates during the 2017 and 2018 growing seasons. These results are consistent with the results of Tukey's test conducted on biomass data in the previous section, which greater variability among the varieties at the end of each growing season.

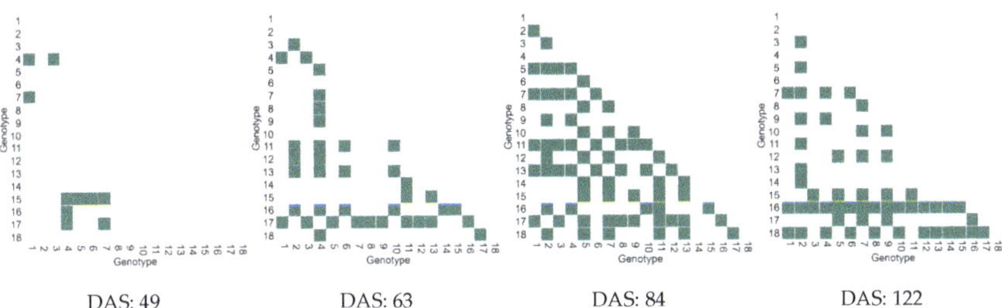

Figure 11. Multiple-comparison Tukey's test for the OSAVI index for the 18 genotypes planted in HyCal-17 experiment collected throughout the 2017 growing season. Green shows the two varieties are significantly different from each other (α = 0.05).

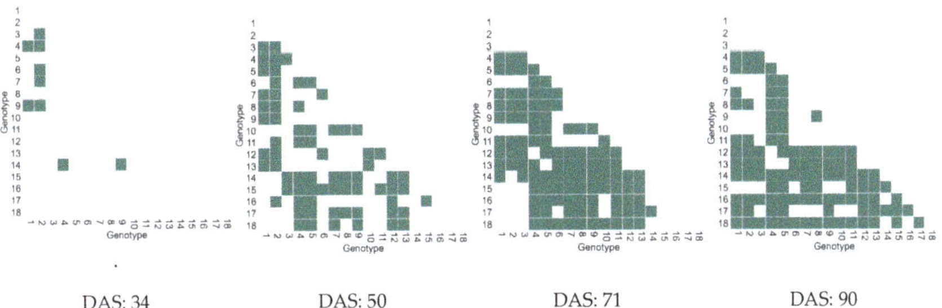

Figure 12. Multiple-comparison Tukey's test for the LiDAR-based volume for the 18 genotypes planted in HyCal-18 experiment collected throughout the 2018 growing season. Green shows the two varieties are significantly different from each other (α = 0.05).

4.1.3. Relationship between Features and Biomass

In this section, the relationship between the biomass data and RS features, as well as the change in their relationship during the growing season, is discussed. Figure 13 shows one feature from LiDAR (90th percentile height of the plants), one feature from hyperspectral data (Intg_NIR1), and the biomass data for the dwarf grain sorghum 341 × 10 and the photoperiod sensitive Trudan Headless varieties, both from the HyCal-17 and HyCal-18 experiments (one with low and one with high biomass values). To compare the data from the two years at the same stage of growth, the data are plotted versus growing degree day (GDD), a heat index calculated from temperature data for each day [68]. At GDD of 2100, the biomass in 2018 for both varieties was slightly higher than in 2017 (as noted in Table 1, the HyCal-18 was planted two weeks earlier than HyCal-17). The height data, Intg_NIR1, and biomass data follow the same pattern of change over time for each variety in both growing seasons. The height for the photoperiod sensitive variety ("Sordan Headless") always increases, while other varieties stop growing around flowering time (GDD of 1500); the Intg_NIR1 increases rapidly earlier in the season, and then gradually decreases at around GDD of 1500 until the end of the season; the biomass continues to increase, and especially for the photoperiod sensitive variety. Inter-annual differences also inherently include the impact of the timing and quantity of rainfall.

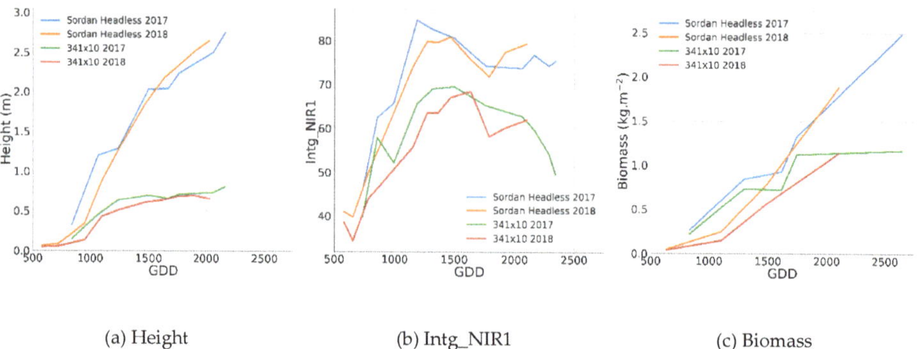

(a) Height (b) Intg_NIR1 (c) Biomass

Figure 13. Comparison of the 90th percentile height (**a**), Intg_NIR1 (**b**), and the biomass data (**c**) for the dwarf grain sorghum 341 × 10 and the photoperiod sensitive Trudan Headless varieties in the 2017 and 2018 growing seasons.

For each feature extracted from the RS data, the simple prediction potential (R^2) and associated changes during a season were investigated using linear regression-based models for the end of season biomass prediction. Robust features should be applicable across the varieties, at least for common experiments. Figures 14 and 15 show the R^2 values of the models for each feature extracted from LiDAR and hyperspectral VNIR data, at four stages of growth and for all nine experiments conducted in the 2017 and 2018 growing seasons. From Figure 14, the 30th percentile height and volume features provided the highest R^2 values for predicting the end of season biomass among the LiDAR-based features for both HyCal-17 and HyCal-18 experiments, as the varieties in those experiments were more diverse in their structural characteristics, providing strong potential for biomass prediction using geometric-related features. The R^2 values for different LiDAR-based features in the InCal-17, InCal-18, and SbDiv-17 experiments are very similar, which is consistent with Figure 9; they all have the lowest average height and lowest variability in height compared to the other experiments, resulting in lower R^2 values for these experiments compared to the experiments with hybrid cultivars. For both InCal-17 and InCal-18 experiments, the highest R^2 values were obtained from feature #5 (coefficient of variation of height), which is representative of the distribution of the points in the canopy point cloud. The R^2 values of the models developed for the SbDivTc-18 and InCalTc-18 are lower than the HyCal-17 and HyCal-18 experiments, but the same features (features #1, #2, and #5) provided the maximum R^2 for all of these experiments, which include hybrid cultivars. The SbBAP-17 also includes hybrid cultivars; however, the R^2 values for all the features are lower compared to all other experiments, mainly because the last LiDAR data were collected on August 30th, and included many photoperiod sensitive cultivars which grew until the final biomass data were collected at the harvest (28 September). Other varieties did not grow during this time, which impacted the biomass–height relationships. Generally, for the experiments with the hybrid cultivars (refer to Table 1), the late season data sets provided the highest R^2, while for the experiments that included inbred cultivars, the data sets of GDDs yielded the lowest R^2 values.

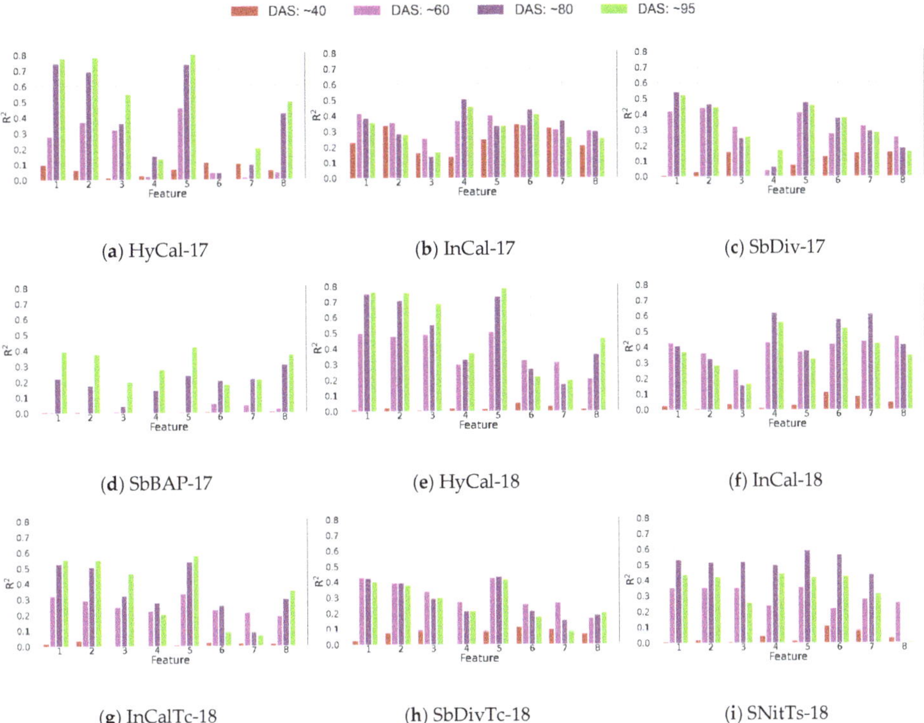

Figure 14. R^2 values of the linear regression-based models developed for the end of season fresh biomass using LiDAR-based features at four stages of growth. Features 1 to 8 represent: #1: 30th percentile height, #2: 50th percentile height, #3: 95th percentile height, #4: coefficient of variation of height, #5: volume, #6: canopy cover (threshold = 0.1), #7: canopy cover (threshold = 0.3), #8: canopy cover (threshold = 0.5).

For the hyperspectral features shown in Figure 15, the highest R^2 values are associated with InCal-17 and InCal-18, collected on ~80 DAS, while for other experiments, the dataset of ~95 DAS yielded the highest R^2 values. Moreover, the same pattern for R^2 values for the features of the InCal-17, InCal-18, and SbDiv-17 was observed. For these panels, the R^2 is generally higher than the panels that include hybrid cultivars.

Given similar trends of the regression models shown in Figures 14 and 15, the models were developed across all the experiments for each of the features, and all the available dates to investigate the potential for using a common set of features for all experiments and times for the multiple input predictive models. The average R^2 for each feature, from all the dates and all the experiments is provided in Figure 16, which shows volume, 30th percentile height, OSAVI, FDR-min, and NDWI features had the highest average R^2 from LiDAR, VNIR, and SWIR data sets, respectively. Linear regression models were also developed for the individual band values from both hyperspectral VNIR and SWIR data. The average and maximum R^2 for each band from all the dates and all the experiments are shown in Figure 17, which shows that the area of spectrum between 750 and 1100 nm provided the highest R^2 for the linear regression models. While the R^2 values for some experiments and some dates for the bands in 2000–2300 nm are relatively high (30–60%), the average R^2 values are much lower in comparison to the 750–1100 nm range.

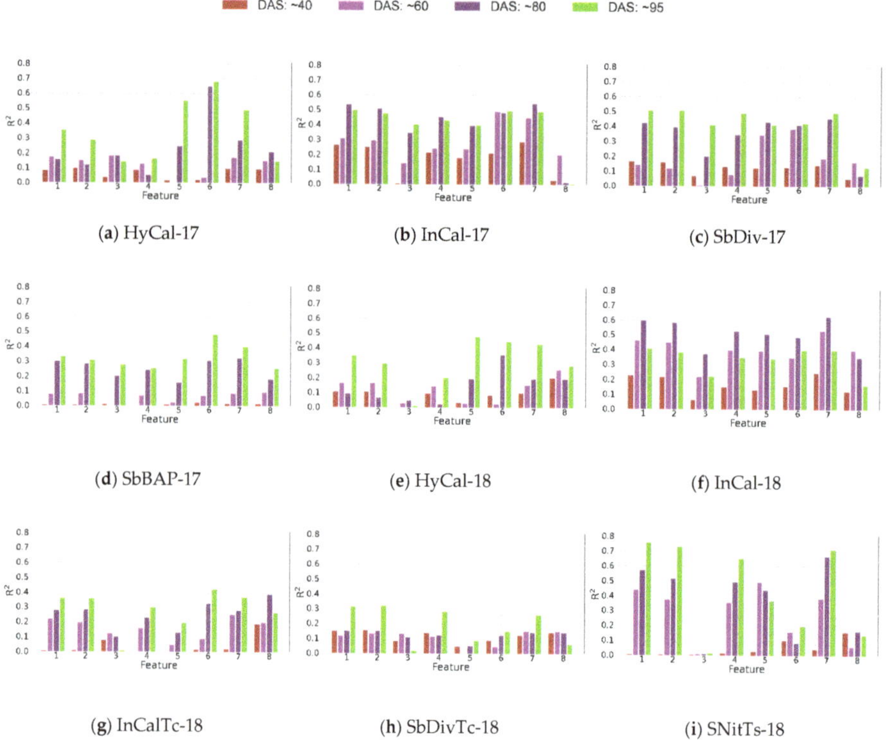

Figure 15. R^2 values of the linear regression-based models developed for the end of season fresh biomass using visible and near infrared (VNIR) features at four stages of growth. Feature 1 to 8 represent: #1: FDR_min, #2: Intg_NIR1, #3: SDR_slope, #4: Intg_NIR1, #5: NDVI, #6: $SR_{800,680}$, #7: OSAVI, and #8: MCARI.

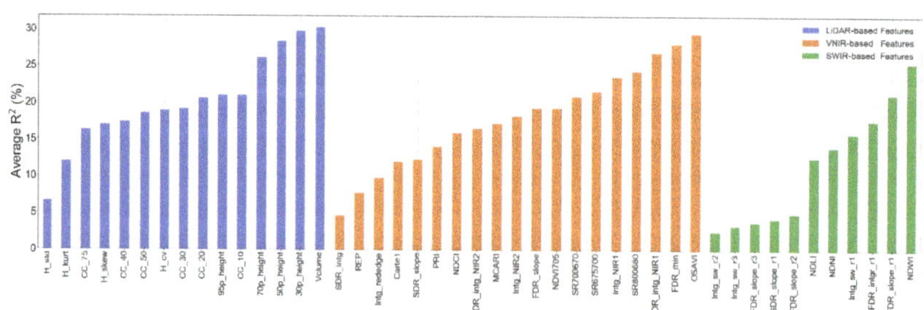

Figure 16. The average R^2 values of the linear models developed for all the dates and all the experiments for each feature type from hyperspectral and LiDAR data.

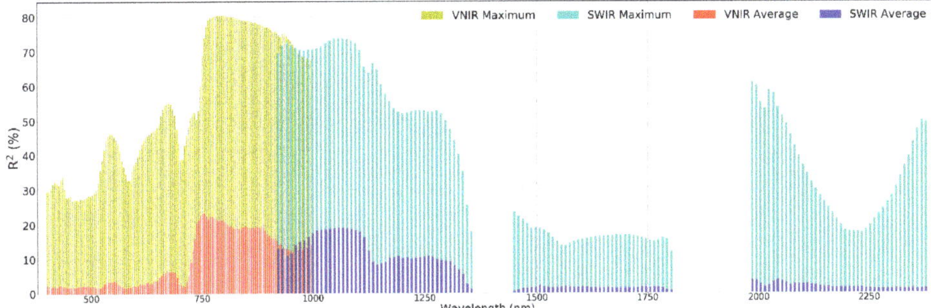

Figure 17. Average and maximum R^2 values of the linear models developed for all the dates and all the experiments for each band from hyperspectral VNIR and short-wave infrared (SWIR) data.

4.2. Biomass Predictive Models

In this section, the results related to the impact of different regression methods, the time of biomass sampling and remote sensing data acquisition, and the number of samples on the prediction results are provided.

4.2.1. Impact of the Data Source and Regression Method on the Prediction Results

To evaluate the performance of different regression-based modeling approaches, PLSR, SVR, and RF were implemented for end of season biomass prediction using the LiDAR and hyperspectral data collected in each growing season. Figure 18 shows the R^2 values of the predictions relative to the reference data for all the experiments, using six data sources (LiDAR, VNIR, SWIR, and combinations), and the three methods. For each prediction with a data source, all available data sets over the whole season were used for training and validation of the models, where two thirds of the sample data (or a maximum of 200 samples) were randomly selected 100 times for the training of the algorithm, and the remaining samples were used for cross validation via the hold-out method. For all the experiments except the SNitTs-18, all the replicates of a variety were assigned to either the training or test sets to avoid any impact from the number of replicates on the prediction results. SNitTs-18, however, included only four varieties, and a different number of replicates for each variety; thus, the training and test sets were assigned randomly from the plots regardless of their varieties for this experiment. Potential reasons for differences in predictions include:

(i) Diversity in the samples: the regression models are better able to learn the pattern in the data when the samples are more diverse.
(ii) Number of data samples: the larger the number of data points in an experiment, the higher accuracies are typically achieved for the prediction.
(iii) Similarity between the training and test data sets; if the training and test data sets are very different, then overfitting can occur for the training data set, resulting in decreased accuracy of the predictions. Note that this can happen when the number of data samples is limited, which causes unlike training and test sets, even when the samples are selected randomly. Additionally, if there is a significant range of biomass values in one experiment, there is more chance to have dissimilar training and test sets.

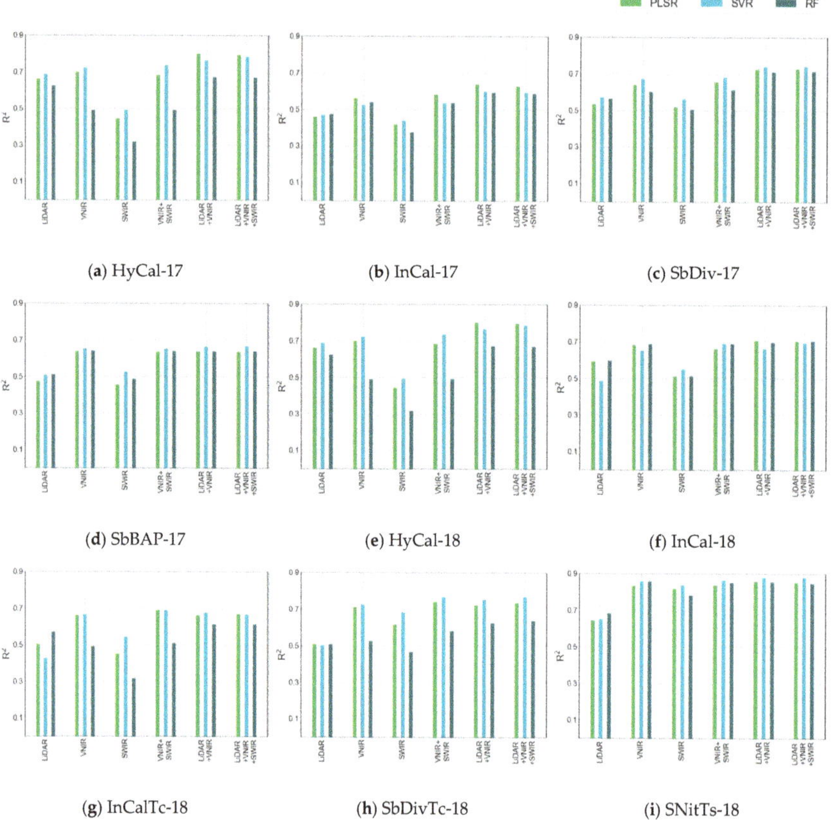

(**a**) HyCal-17 (**b**) InCal-17 (**c**) SbDiv-17

(**d**) SbBAP-17 (**e**) HyCal-18 (**f**) InCal-18

(**g**) InCalTc-18 (**h**) SbDivTc-18 (**i**) SNitTs-18

Figure 18. R^2 values of the end of season fresh biomass predictions, using six data sources, and the three partial least squares regression (PLSR), support vector regression (SVR), and Random Forest (RF) methods for all the experiments conducted in the 2017 and 2018 growing seasons.

Figure 18 shows that the highest accuracy of end of season biomass prediction using all combinations of data sources was achieved for the SNitTs-18 experiment. There were two nitrogen treatments in this experiment; half of the plots in this experiment were fertilized with 250 kg/ha nitrogen while the other half were not fertilized, causing high and low biomass values for the plots, high diversity in the reflectance data from the hyperspectral images, as well as high diversity in geometric-based features extracted from LiDAR point cloud (reason i). As was noted, samples were assigned to the training and test sets for this experiment differently from the other experiments, causing multiple samples of each variety to be assigned to both the training and test sets, resulting in increased similarity in the two sets (reason iii).

The highest accuracy of prediction using LiDAR features as the sole input was obtained for the HyCal-17 and HyCal-18 experiments, which include hybrid cultivars that are more diverse in structural characteristics compared to the inbred cultivars; thus, the regression model can distinguish and relate the LiDAR-based features to the biomass data (reason i), which is consistent with the results in Figure 14. In general, the predictions are more accurate for the experiments that include hybrid cultivars. As was shown in Figure 9, the InCal-17, InCal-18, and SbDiv-17 have the smallest standard deviation in the LiDAR-based height, indicating that the associated varieties have similar structural characteristics.

The predictions for the SbDiv-17 are more accurate than the predictions for the InCal-17 as more samples were available for the training set, 200 for SbDiv-17, and 80 for InCal-17 (reason ii). For the SbBAP-17, the prediction accuracies are lower than most of the other experiments. This experiment included varieties that were highly diverse in terms of structural characteristics (Figure 9), also had a much larger range compared to the other experiments. This resulted in dissimilar samples in the training and test sets (reason iii). Figure 19 shows a box plot for the fresh biomass data for all the experiments, and the SbBAP-17 experiment had the greatest range of biomass values among the experiments, the lowest accuracies for the predictions, while SNitTs-18 experiment has the smallest range and the highest R^2 values for the predictions.

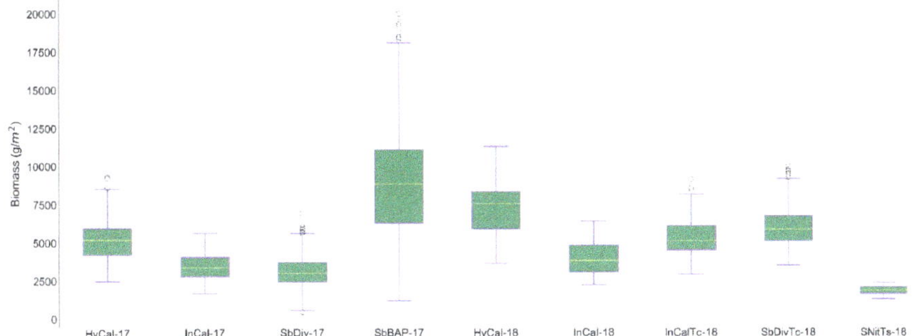

Figure 19. The box plot for the fresh biomass data for all the experiments conducted in the 2017 and 2018 growing seasons.

For HyCal-17 and HyCal-18 experiments, PLSR, SVR, and RF models were developed using all three data sources and leave-one-out cross validation strategy, where in each fold, one variety was assigned as test and the other 17 varieties were included in the training set. The results are shown in Figure 20. For both experiments, the SVR method provided the highest R^2 values for the predictions. For HyCal-17, all three regression methods underestimated the value of the biomass for one of the photoperiod varieties (which also had the maximum biomass, as noted previously); however, the RF model had the lowest accuracy. RF for both years resulted the lowest R^2 as a result of overfitting as was discussed earlier. All three methods resulted in predictions with lower accuracies for the experiment in 2018 compared to 2017, which could be because the end of season biomass data were measured at an earlier date in 2018, when all the plants had not reached full maturity.

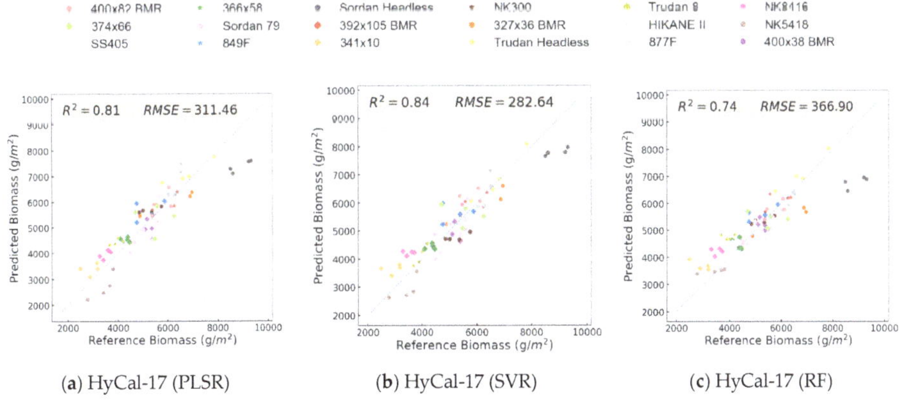

(**a**) HyCal-17 (PLSR) (**b**) HyCal-17 (SVR) (**c**) HyCal-17 (RF)

Figure 20. *Cont.*

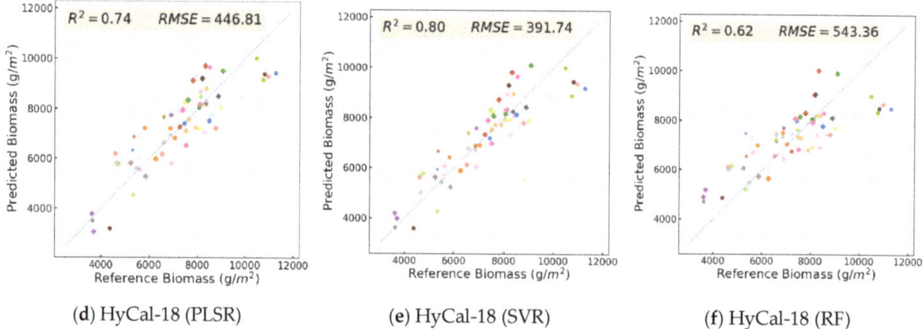

(**d**) HyCal-18 (PLSR) (**e**) HyCal-18 (SVR) (**f**) HyCal-18 (RF)

Figure 20. Prediction results of PLSR, SVR, and RF models developed for the HyCal-17 and HyCal-18 experiments, using all the data sources and leaving-one-out cross validation strategy.

4.2.2. Predictions in Time

To evaluate the capability of remotely sensed data for predicting biomass through the growing season, SVR models were developed for six dates in the 2017 and 2018 growing season for the HyCal-17 and HyCal-18 experiments. The R^2 values of the predictions relative to the reference data are shown in Figure 21. For each prediction, all the VNIR hyperspectral and LiDAR data collected prior to the date of biomass measurement were used in the SVR models. The R^2 values of the predictions at the beginning of the season were lower compared to the end of the season, especially when using only VNIR features. Early season growth is focused on the production of biomass from stalks and leaves, while mid-season development is related to flowering and early development of panicles. Plant structural characteristics do not change significantly after flowering in the mid-season, while spectral characteristics change significantly especially during flowering with the emergence of the panicles.

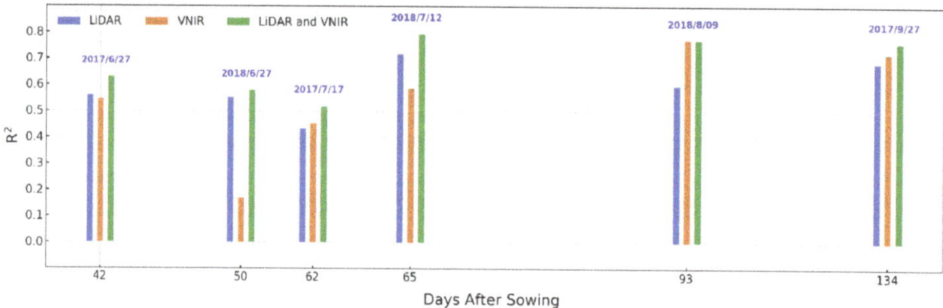

Figure 21. The R^2 of predicted biomass during the 2017 and 2018 growing seasons using the SVR models developed based on VNIR hyperspectral and LiDAR data for the hybrid calibration panels.

4.2.3. Multi-Temporal Predictions of End-of-Season Biomass

It is highly desirable to predict the end-of-season biomass as early as possible during the growing season to avoid unnecessary investment of phenotyping resources in non-productive varieties. The SbBAP-17, SbDiv-17, and SbDiv-18 were chosen to conduct the evaluations in this section as they had an adequate number of samples as well as RS data points and included both hybrid and inbred cultivars. Figure 22 shows the accuracy of end-of-season biomass predictions for these experiments using hyperspectral VNIR, SWIR, and LiDAR data from each date individually, and in combination with the earlier dates. For both SbBAP-17 and SbDiv-17 experiments, the earliest data set yielded very low prediction accuracies. For SbBAP-17, the best results when using features from individual sensors for VNIR and SWIR data sets were achieved from 10 September with $R^2 = 0.60$ and $R^2 = 0.54$, respectively. Based on LiDAR data, 23 August resulted in the highest values, with $R^2 = 0.46$. For the SbDiv-17 experiment, the combined VNIR and LiDAR data sets of 25 July and 2 August provided the highest accuracy of using individual data sets. For SbDivTc-18, the data inputs from 11 July resulted an R^2 of 0.75, which indicates the July data sets have good potential for biomass predictions. For all three experiments, the best results were obtained when features from all the hyperspectral and LiDAR data sets from the whole season were used, resulting in R^2 of 0.63, 0.75, and 0.78 for the SbBAP-17, SbDiv-17, and SbDiv-18 experiments, respectively. Although the best results were obtained using the whole season RS data, the models developed using middle season data (DAS of ~60 to 80) were also able to provide comparable accuracies.

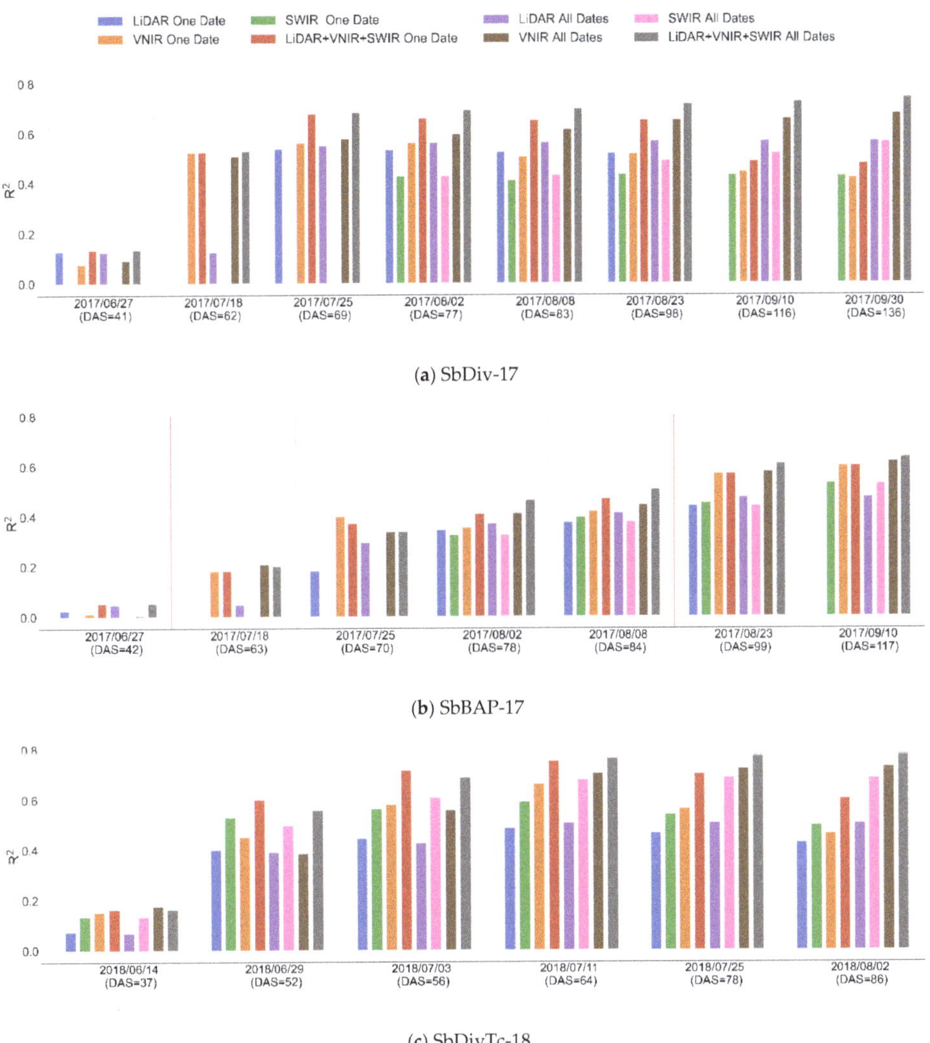

(**a**) SbDiv-17

(**b**) SbBAP-17

(**c**) SbDivTc-18

Figure 22. R^2 for end of season predictions using hyperspectral and LiDAR data collected on different dates for the SbBAP-17, SbDiv-17, and SbDivTc-18 experiment.

4.2.4. Impact of the Number of Features and Samples on Biomass Prediction

As noted earlier, measuring biomass in the field is time-consuming and expensive; however, it is still required for training the regression models. Generally, the greater the number of samples for training, the more accurate the predictions. To evaluate the impact of the number of samples in training set, the SVR and PLSR models are developed for end-of-season biomass prediction using all the data sources and various numbers of samples in the training set. Each model was trained with a specific number of samples, and the process was repeated 100 times, each time with a different, but same sized set of randomly selected samples. The rest of the available samples were assigned to the testing set. Figure 23 shows the median and standard deviation of the R^2 values of developed models for some of the experimental trials. The R^2 of predictions for both SVR and PLSR models increase as the number of training samples increases. However, the accuracy of PLSR models is higher than SVR models when a smaller number of training samples is used. The rate of increase of R^2 with the respective increases in the number of training samples is higher for SVR compared to PLSR; thus, when the maximum of available samples is used in training for experiments (e.g., for SbDivTc-18), the SVR models had higher R^2 values. For all the experiments, the standard deviation of the R^2 values decreases as the number of training samples increases, showing more reliable (repeatable) prediction models are developed when more samples are available for training, as expected. However, for some experiments such as HyCal-17, the standard deviation of R^2 decreases initially, reaches a minimum, then increases. This is attributed to the small total number of samples: using more samples in the training set implies a smaller number of samples is available in the test set.

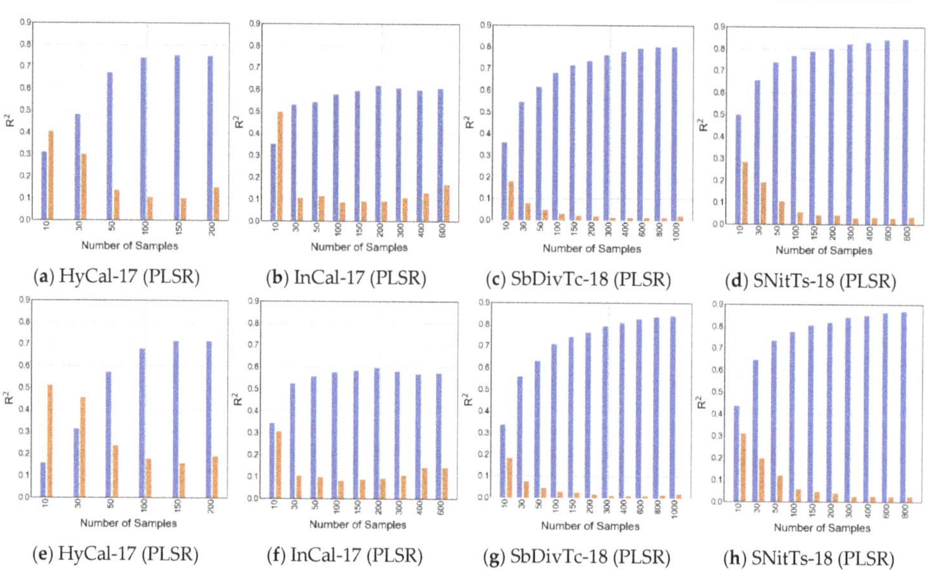

(**a**) HyCal-17 (PLSR) (**b**) InCal-17 (PLSR) (**c**) SbDivTc-18 (PLSR) (**d**) SNitTs-18 (PLSR)

(**e**) HyCal-17 (PLSR) (**f**) InCal-17 (PLSR) (**g**) SbDivTc-18 (PLSR) (**h**) SNitTs-18 (PLSR)

Figure 23. Impact of the number of samples on R^2 of the predictive models using SVR and PLSR models.

5. Discussion

The predictions results with respect to the diversity, number of samples, and similarity between test and training sets for all the experiments are summarized in Table 3.

Table 3. Summary of the prediction results of the experimental trials.

	HyCal-17 and 18	InCal-17 and 18	SbBAP-17	SbDiv-17	InCalTc-18	SbDivTc-18	SNitTs
Diversity	Very High	Low	Very High	Low	High	High	High
Number of samples	Low	Low	High	High	Low	High	Low
Test-training set dissimilarity (range of biomass)	High	Low	Very High	Low	High	High	Low
Prediction accuracy (maximum R^2)	High (0.80)	Medium (0.71)	Medium (0.67)	Medium (0.74)	Medium (0.69)	High (0.77)	Very high (0.88)

In general, for the experiments with hybrid cultivars, RF models had lowest prediction accuracies among the three methods, which is related to the fact that there was more dissimilarity between the training and test sets in both RS and biomass data among the hybrid cultivars compared to those that were inbred, and RF models can be overfitted to the training data set; thus, they may not provide as accurate predictions as SVR and PLSR. For the InCal-18, however, RF yielded the highest accuracies for most of the data sources. For the experiments with a sample size of 200 (SbDiv-17, SbBAP-17, and SbDivTc-18), the SVR models provided the most accurate results, while for the experiments with a lower number of data samples, PLSR provided the highest prediction accuracies.

A summary of the prediction results for various data sources and regression methods is provided in Figure 24, where the R^2 values of the nine experiments are shown in a box plot (RMSE values are shown in Figure A6 in Appendix B). For the LiDAR data, the RF method provided slightly higher median accuracies than PLSR and SVR, with lower variability in R^2 values (more reliability). When VNIR data was the only input, PLSR yielded more accurate results, which is similar to the results obtained in [10] yield prediction of potatoes using VNIR hyperspectral data. For all other data sources, SVR yielded a higher median R^2. For SWIR and VNIR combined with SWIR sources, SVR provided more reliable results, while for VNIR, VNIR combined with LiDAR, as well as a combination of all data sources, PLSR provided more reliable results. Also, the SVR models provided the maximum R^2 for all the data sources.

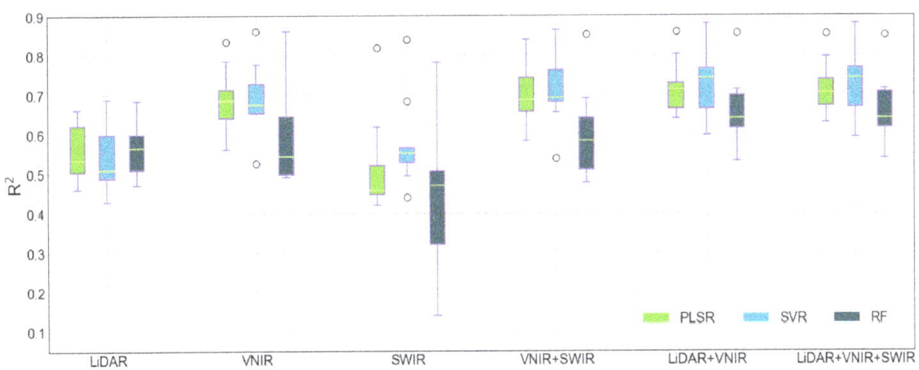

Figure 24. Box plot of the prediction results for various data sources and regression methods.

Similar to the study by Almeida et al. [69], an ANOVA test was performed on the prediction results to determine the impact of the method and data source (e.g., sensor-based features) on the prediction results. The ANOVA results provided in Table 4 show that the data source is the cause of 69% of the variation in the prediction results, 28% for the regression method, and 3% for the interaction their interaction. These results are consistent with those of [33,69]. This indicates that the data source should also be considered to determine the regression method in the design of similar experiments. A similar ANOVA test was performed on the R^2 values for six experiments, including HyCal-17, HyCal-18,

InCal-17, InCal-18, SbDiv-17, and SbDiv-18, of which three include hybrid cultivars, and three include inbred cultivars. On this test, the cultivar type (hybrid or inbred) was considered as the third factor. Recall that the sorghum hybrid cultivars are more variable in their characteristics than inbreds in terms of biomass and structural characteristics. The results of this test in Table 5 indicate that the data source also has the highest contribution to the variation in the predictions (44%). The regression method is the cause of 26% of the variation in the prediction results, and less than 1% is attributed to the cultivar type. However, the interaction between the regression method and cultivar is responsible for 24% of the variation, suggesting that the cultivar type is another important factor to consider when determining the regression method for developing predictive models.

Table 4. Analysis of variance of the R^2 respective to the data source, regression method, and their interaction.

Factor	Sum of Squares	Degree of Freedom	F Value	p-Value	η^2 (%)
Data source	103.70	5	564.09	$<2 \times 10^{-14}$	68.59
Method	17.15	2	233.23	$<2 \times 10^{-14}$	28.36
Data source:method	9.24	10	25.14	$<2 \times 10^{-14}$	3.06
Residuals	594.95	16,182			

Table 5. Analysis of variance of the R^2 respective to the data source, regression method, cultivar type, and their interactions.

Factor	Sum of Squares	Degree of Freedom	F Value	p-Value	η^2 (%)
Data source	80.95	5	412.58	$<2 \times 10^{-6}$	44.18
Method	19.29	2	245.79	$<2 \times 10^{-6}$	26.32
Cultivar type	0.35	1	8.86	0.003	0.95
Data source:method	6.84	10	17.44	$<2 \times 10^{-6}$	1.87
Data source:cultivar type	4.03	5	20.53	$<2 \times 10^{-6}$	2.20
Method:cultivar type	17.59	2	224.12	$<2 \times 10^{-6}$	24.00
Data source:method:cultivar type	1.82	10	4.63	0.093	0.50
Residuals	422.37	10,764			

Recommendations for Biomass Prediction

In this section, we summarize the findings of the tests conducted in this paper in the format of recommendations to achieve reliable biomass prediction. The recommendations are based on the data, the genotypes, and the location where the study was conducted; thus, they might not be generalizable.

- Regression Model

Both PLSR and SVR models provided more accurate predictions than RF. SVR generally provided more accurate predictions, however, PLSR is preferable when the number of sample data points is very limited (less than 50 samples) as well as when high variability in the biomass data is expected.

- Data Source

ANOVA analysis on the prediction results showed that the data source is the most important factor on determining the accuracy of the predictions. Considering individual sensors, the features extracted from VNIR provided the most accurate predictions. However, adding the geometric based features extracted from the LiDAR data to the models improved the accuracy of the predictions significantly, which is consistent with previous studies for biomass prediction in forest environments [44,69]. We recommend acquiring data from both VNIR and LiDAR sensors, if possible, for the most reliable biomass prediction.

- Flight Time and Frequency

The analysis on correlation between the data sets captured throughout the season showed that changes occur more rapidly earlier in the season (a result of fast-growing plants), while the multi-temporal data captured later in season are more similar. This implies that frequent data collection at the end of season is not required. Moreover, the most accurate end of season biomass prediction was achieved using the data captured around 60–80 DAS, which can be considered as an important time for collecting RS data. Zhou et al. [70] obtained similar results, but for rice yield prediction using UAV-based multispectral and digital imagery.

- Required Reference Data

Based on the results provided in Figure 24, we recommend collecting least 50 samples. If it is expected to have high variability in the biomass data associated with the varieties in the experiments, more samples would be required.

6. Conclusions

In this paper, we explored the potential for reliable prediction of sorghum biomass using multi-temporal hyperspectral and LiDAR data acquired by sensors mounted on UAV platforms. We developed prediction models using three nonlinear regression models for nine experiments conducted in the 2017 and 2018 growing seasons at the Agronomy Center for Research and Education (ACRE) at Purdue University. Experiments included multiple sorghum varieties with different sample sizes, providing an opportunity for multiple statistical tests and models. Based on the experiments conducted in this study, nitrogen and photosynthesis related features extracted from hyperspectral data and geometric based features derived from the LiDAR data provided reliable and accurate prediction of biomass. The 750–1100 nm range of the spectrum provided the most relevant information for biomass prediction.

Both data source and regression method are important factors for a reliable prediction; however, the ANOVA results show that the data source was more important with 69% significance, versus 28% significance for the regression method. The number of samples in training set for the prediction is an important factor for determining the accuracy of the predictions. Generally, the PLSR method provided more accurate prediction models when the number of samples in training was limited. With increasing samples, the rate of increase in the accuracy of the SVR models was higher than PLSR.

We also evaluated the prediction models with respect to the time of the RS data acquisition and the time of harvest. The end-of-season biomass predictions were more reliable and accurate than the mid-season predictions, as more varieties in the field were at the same stage of growth. With respect to the remote sensing data, the best results were obtained using the RS data from the whole season; however, the models developed using mid-season data (DAS of ~60 to 80) were also able to provide comparably accurate results, which were useful for early screening of varieties.

Author Contributions: A.M. conducted all the statistical analyses including all the figures and wrote the manuscript. N.R.C., M.M.C., and M.R.T. provided essential help and feedback on the paper content. M.M.C. supervised the acquisition of RS data. A.M., N.R.C., M.M.C., and M.R.T. interpreted the results. N.R.C. and M.R.T. designed the experiments and analyzed the phenotypic and marker data. All authors have read and agreed to the published version of the manuscript.

Funding: This research was funded by the Advanced Research Projects Agency-Energy (ARPA-E), U.S. Department of Energy under Grant DE-AR0000593.

Acknowledgments: The authors thank the Purdue TERRA team; Evan Flatt for his work on system integration, flying, and data processing; Addie Thompson, Kai-Wei Yang, and Andrew Linvill for their contributions to collection, processing, and finalizing the ground reference data; Meghdad Hasheminasab, Tian Zhao, Magdy Elbahnasawy, Tamer Shamseldin, Radhika Ravi, Yun-Jou Lin, and Yi Chun Lin for their work on collecting and processing the RGB and LiDAR data; Karoll Quijano, Ruya Xu, Taojun Wang, Behrokh Nazeri, and Zhou Zhang for their work on hyperspectral data collection and processing; Professor Ayman Habib for his valuable input throughout this work.

Conflicts of Interest: The authors declare no conflict of interest.

Appendix A

Table A1. Commercial varieties planted in the hybrid calibration panel.

Variety	Variety Name	Sorghum Type	Company
1	849F	Forage	Pioneer
2	877F	Forage	Pioneer
3	327 × 36 BMR	Forage	Richardson
4	341 × 10	Forage	Richardson
5	366 × 58	Food grain	Richardson
6	374 × 66	Food grain	Richardson
7	392 × 105 BMR	Forage	Richardson
8	400 × 38 BMR	Sudangrass	Richardson
9	400 × 82 BMR	Sudangrass	Richardson
10	HIKANE II	Forage	Sorghum Partners
11	NK300	Forage	Sorghum Partners
12	NK5418	Grain	Sorghum Partners
13	NK8416	Grain	Sorghum Partners
14	SS405	Forage	Sorghum Partners
15	Sordan 79	Forage	Sorghum Partners
16	Sordan Headless	Forage (photoperiod sensitive)	Sorghum Partners
17	Trudan 8	Forage	Sorghum Partners
18	Trudan Headless	Forage (photoperiod sensitive)	Sorghum Partners

Table A2. Remote Sensing Data Sets.

Year	Data Type	Field	Dates
2017	RGB and LiDAR	InCal, HyCal, SbDiv, and SbBAP	16/06, 21/06, 27/06, 05/07, 11/07, 14/07, 17/07, 25/07, 02/08, 08/08, 23/08, 3008
	VNIR	HyCal and InCal	21/06, 28/06, 04/07, 12/07, 18/07, 25/07, 08/08, 23/08, 30/08, 10/09, 15/09
		SbDiv	21/06, 27/06, 04/07, 18/07, 25/07, 30/07, 08/08, 14/08, 23/08, 10/09, 24/09, 30/09
		SbBAP	21/06, 27/06, 04/07, 18/07, 25/07, 30/07, 08/08, 14/08, 23/08, 10/09, 24/09
	SWIR	InCal and HyCal	23/08, 30/08, 10/09, 15/09
		SbDiv	02/08, 08/08, 14/08, 23/08, 30/08, 10/09, 30/09
		SbBAP	02/08, 08/08, 14/08, 23/08, 30/08, 10/09
2018	RGB and LiDAR	HyCal, InCal, InCalTc, and SbDivTc	22/05, 29/05, 04/05, 11/06, 20/06, 27/06, 02/07, 11/07, 18/07, 23/07, 01/08, 06/08
		SNitTs	28/06, 03/07, 11/07, 17/17, 23/07, 01/08, 06/08, 16/08, 25/08, 05/09, 19/09
	VNIR and SWIR	HyCal, InCal, InCalTc	04/06, 08/06, 14/06, 29/06, 03/07, 06/07, 11/07, 18/07, 25/07, 02/08, 09/08
		SbDivTc	04/06, 08/06, 14/06, 29/06, 03/07, 06/07, 10/07, 11/07, 25/07, 02/08
		SNitTs	28/06, 03/07, 11/07, 18/07, 25/07, 02/08, 13/08, 28/08, 04/09, 12/09, 18/09

Table A3. Vegetation indices extracted from each HSI spectrum.

Data Type	Index Name	Formulation	References
VNIR	NDVI	$(R_{750} - R_{705})/(R_{750} + R_{705})$	[70]
	NDCI	$(R_{762} - R_{527})/(R_{762} + R_{527})$	[71]
	Carte1	R_{695}/R_{420}	[72]
	$SR_{800,680}$	R_{800}/R_{680}	[73]
	$SR_{675,700}$	R_{675}/R_{700}	[74]
	$SR_{700,670}$	R_{700}/R_{670}	[75]
	OSAVI	$(1 + 0.16)\ (R_{800} - R_{670})/(R_{800} + R_{670} + 0.16)$	[76]
	MCARI	$[(R_{700} - R_{670}) - 0.2(R_{700}\text{-}R_{550})](R_{700}/R_{670})$	[77]
	REP	$700 + 40[(R_{670} + R_{780})/2 - R_{700})]/(R_{740} - R_{700}))$	[78]
	PRI	$(R_{531} - R_{570})/(R_{531} + R_{570})$	[79]
SWIR	NDWI	$(R_{860} - R_{1240})/(R_{860} + R_{1240})$	[80]
	NDLI	$[\log(1/R_{1754}) - \log(1/R_{1680})]/[\log(1/R_{1754}) + \log(1/R_{1680})]$	[81]
	NDNI	$[\log(1/R_{1510}) - \log(1/R_{1680})]/[\log(1/R_{1510}) + \log(1/R_{1680})]$	[81]

Table A4. Integration features extracted from each HSI spectrum.

Data Type	Feature Name	λ_a	λ_b
VNIR	Intg_rededge	685	745
	Intg_NIR1	770	910
	Intg_NIR2	910	1000
SWIR	Intg_SWIR_r1	920	1353
	Intg_SWIR_r2	1430	1800
	Intg_SWIR_r3	1952	2385

Table A5. Derivative features were extracted from each first derivative (FDR) and second derivative (SDR) spectrum.

Data Type	Feature Name	Description
VNIR	FDR_slope	slope of the line that passes through the minimum of FDR and the maximum of FDR in 660–690 nm range
	FDR_min	minimum of FDR
	FDR_intg_ NIR1	integration of FDR in bands between 670 and 780 nm
	FDR_intg_ NIR2	integration of FDR in bands between 910 and 1000 nm
	SDR_slope_rededge	slope of the line passing through the maximum and the minimum of SDR
	SDR_intg	integration of SDR in all bands
SWIR	FDR_slope_r1	slope of the line that passes through the maximum of FDR in 1000–1050 nm range and the minimum of FDR in 1100–1200 nm range
	FDR_slope_r2	slope of the line that passes through the maximum of FDR in 1475–1525 nm range and the minimum of FDR in 1675–1725 nm range
	FDR_slope_r3	slope of the line that passes through the maximum of FDR in 2000–2050 nm range and the minimum of FDR in 2200–2300 nm range
	FDR_intg-r1	integration of FDR in bands between 920 and 1353 nm
	SDR_ slope_r1	slope of the line that passes through the maximum of SDR in 1100–1200 nm range and the minimum of SDR in 1000–1100 nm range

Table A6. Grid search parameters for regression methods.

Algorithm	Hyperparameter	Values Tested
PLSR	Number of components	2, 3, 5, 10, 15, 20
SVR (RBF)	C gamma	10, 100, 1000, number of features 1/n, 0.0001, 0.001, 0.01, 0.1
Random Forest (RF)	Max tree depth Min sample split Number of trees	5, 10, 100 2, 10 50, 100, 500

Appendix B

(**a**) InCal-17 (left) and HyCal-17 (right)

(**b**) SbBAP-17

(**c**) SbDiv-17

Figure A1. RGB images of the field trials in 2017.

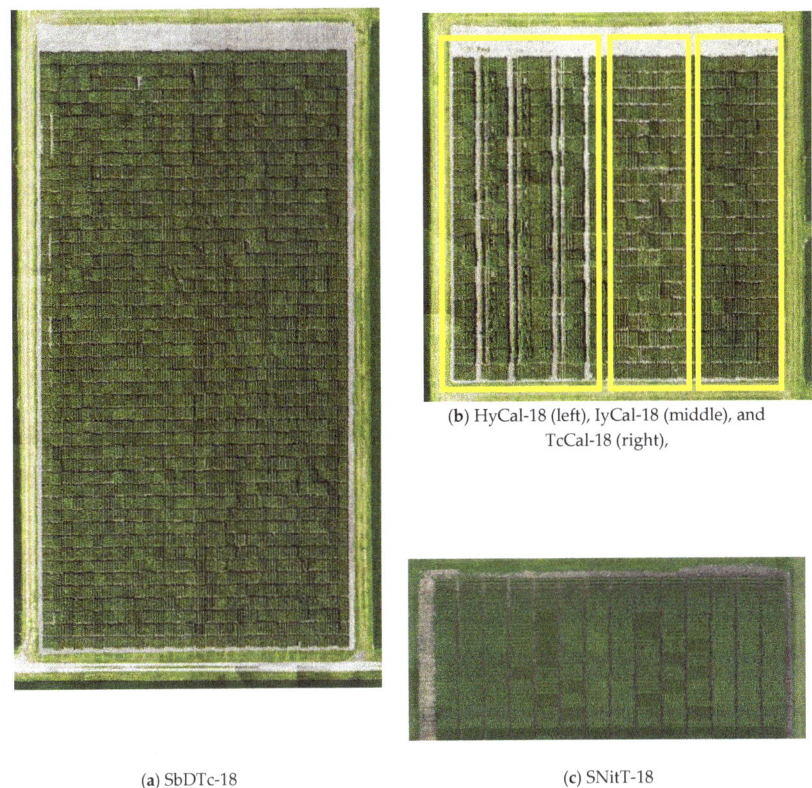

(**b**) HyCal-18 (left), IyCal-18 (middle), and TcCal-18 (right),

(**a**) SbDTc-18

(**c**) SNitT-18

Figure A2. RGB images of the field trials in 2018.

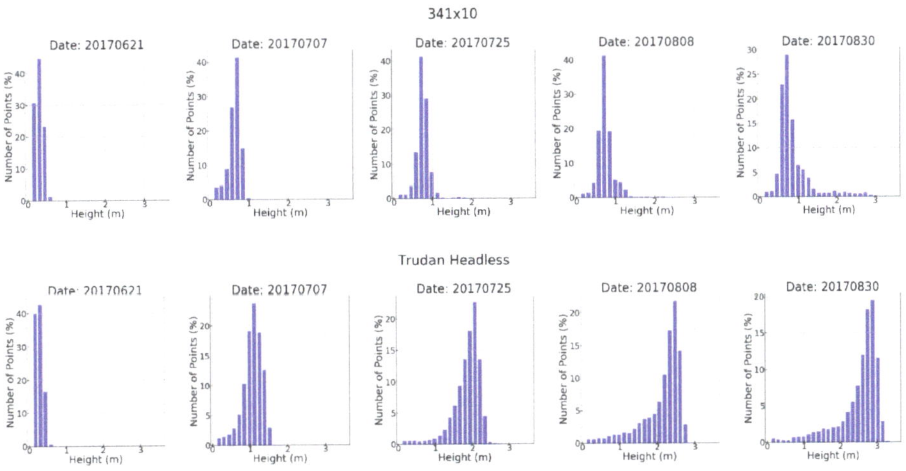

Figure A3. Height histogram for "341 × 10" and "Trudan Headless" in the HyCal panel across the 2017 growing season.

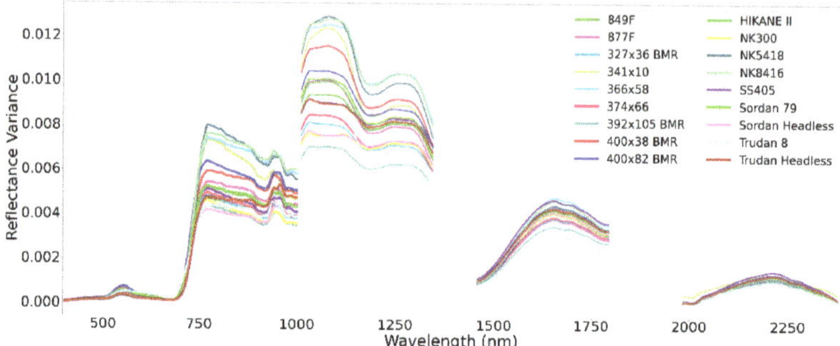

Figure A4. Variance of the spectra of the 18 Sorghum Varieties in HyCal-18 panel on 18 July 2018. The varieties are very similar in the visible range of the spectrum, but substantial variability is observed in the NIR portion of spectrum.

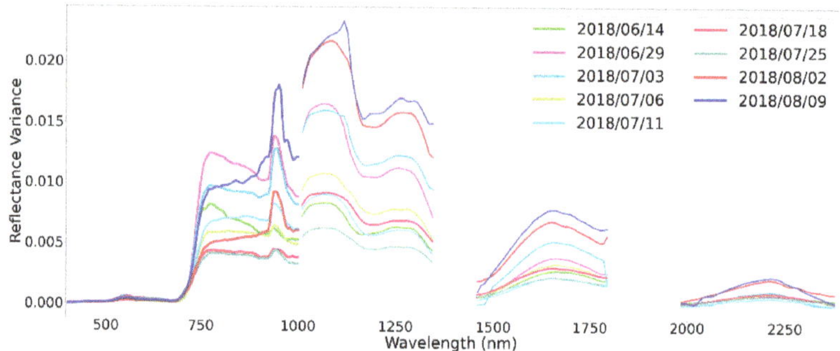

Figure A5. Example of reflectance variance of one of the varieties in HyCal-18 experiment ("Trudan 8") during the 2018 growing season.

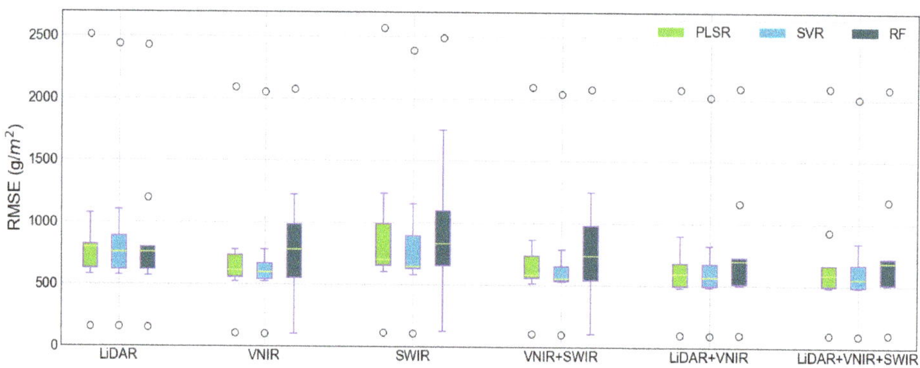

Figure A6. Box plot of the prediction results for various data sources and regression methods.

References

1. Davey, J.W.; Hohenlohe, P.A.; Etter, P.D.; Boone, J.Q.; Catchen, J.M.; Blaxter, M.L. Genome-wide genetic marker discovery and genotyping using next-generation sequencing. *Nat. Rev. Genet.* **2011**, *12*, 499–510. [CrossRef]
2. Potgieter, A.B.; George-Jaeggli, B.; Chapman, S.C.; Laws, K.; Suárez Cadavid, L.A.; Wixted, J.; Watson, J.; Eldridge, M.; Jordan, D.R.; Hammer, G.L. Multi-Spectral Imaging from an Unmanned Aerial Vehicle Enables the Assessment of Seasonal Leaf Area Dynamics of Sorghum Breeding Lines. *Front. Plant Sci.* **2017**, *8*, 1532–1546. [CrossRef] [PubMed]
3. Liang, L.; Di, L.; Zhang, L.; Deng, M.; Qin, Z.; Zhao, S.; Lin, H. Estimation of crop LAI using hyperspectral vegetation indices and a hybrid inversion method. *Remote Sens. Environ.* **2015**, *165*, 123–134. [CrossRef]
4. Chu, T.; Starek, M.J.; Brewer, M.J.; Murray, S.C.; Pruter, L.S. Characterizing canopy height with UAS structure-from-motion photogrammetry—Results analysis of a maize field trial with respect to multiple factors. *Remote Sens. Lett.* **2018**, *9*, 753–762. [CrossRef]
5. Pugh, N.A.; Horne, D.W.; Murray, S.C.; Carvalho, G.; Malambo, L.; Jung, J.; Chang, A.; Maeda, M.; Popescu, S.; Chu, T.; et al. Temporal Estimates of Crop Growth in Sorghum and Maize Breeding Enabled by Unmanned Aerial Systems. *Plant Phenome J.* **2018**, *1*, 1–10. [CrossRef]
6. Maimaitijiang, M.; Ghulam, A.; Sidike, P.; Hartling, S.; Maimaitiyiming, M.; Peterson, K.; Shavers, E.; Fishman, J.; Peterson, J.; Kadam, S.; et al. Unmanned Aerial System (UAS)-based phenotyping of soybean using multi-sensor data fusion and extreme learning machine. *ISPRS J. Photogramm. Remote Sens.* **2017**, *134*, 43–58. [CrossRef]
7. Tattaris, M.; Reynolds, M.P.; Chapman, S.C. A Direct Comparison of Remote Sensing Approaches for High-Throughput Phenotyping in Plant Breeding. *Front. Plant Sci.* **2016**, *7*, 1–9. [CrossRef]
8. Eitel, J.U.H.; Magney, T.S.; Vierling, L.A.; Greaves, H.E.; Zheng, G. An automated method to quantify crop height and calibrate satellite-derived biomass using hypertemporal lidar. *Remote Sens. Environ.* **2016**, *187*, 414–422. [CrossRef]
9. Li, J.; Shi, Y.; Veeranampalayam-Sivakumar, A.N.; Schachtman, D.P. Elucidating sorghum biomass, nitrogen and chlorophyll contents with spectral and morphological traits derived from unmanned aircraft system. *Front. Plant Sci.* **2018**, *9*, 1–12. [CrossRef]
10. Sun, C.; Feng, L.; Zhang, Z.; Ma, Y.; Crosby, T.; Naber, M.; Wang, Y. Prediction of end-of-season tuber yield and tuber set in potatoes using in-season uav-based hyperspectral imagery and machine learning. *Sensors* **2020**, *20*, 5293. [CrossRef]
11. Li, B.; Xu, X.; Zhang, L.; Han, J.; Bian, C.; Li, G.; Liu, J.; Jin, L. Above-ground biomass estimation and yield prediction in potato by using UAV-based RGB and hyperspectral imaging. *ISPRS J. Photogramm. Remote Sens.* **2020**, *162*, 161–172. [CrossRef]
12. Duan, T.; Chapman, S.C.; Guo, Y.; Zheng, B. Dynamic monitoring of NDVI in wheat agronomy and breeding trials using an unmanned aerial vehicle. *Field Crop. Res.* **2017**, *210*, 71–80. [CrossRef]
13. Stanton, C.; Starek, M.J.; Elliott, N.; Brewer, M.; Maeda, M.M.; Chu, T. Unmanned aircraft system-derived crop height and normalized difference vegetation index metrics for sorghum yield and aphid stress assessment. *J. Appl. Remote Sens.* **2017**, *11*, 026035. [CrossRef]
14. Gracia-Romero, A.; Kefauver, S.C.; Fernandez-Gallego, J.A.; Vergara-Díaz, O.; Nieto-Taladriz, M.T.; Araus, J.L. UAV and ground image-based phenotyping: A proof of concept with durum wheat. *Remote Sens.* **2019**, *11*, 1244. [CrossRef]
15. Perich, G.; Hund, A.; Anderegg, J.; Roth, L.; Boer, M.P.; Walter, A.; Liebisch, F.; Aasen, H. Assessment of Multi-Image Unmanned Aerial Vehicle Based High-Throughput Field Phenotyping of Canopy Temperature. *Front. Plant Sci.* **2020**, *11*, 1–17. [CrossRef]
16. Borra-Serrano, I.; Swaef, T.D.; Quataert, P.; Aper, J.; Saleem, A.; Saeys, W.; Somers, B.; Roldán-Ruiz, I.; Lootens, P. Closing the phenotyping gap: High resolution UAV time series for soybean growth analysis provides objective data from field trials. *Remote Sens.* **2020**, *12*, 1644. [CrossRef]
17. Zhang, Z.; Masjedi, A.; Zhao, J.; Crawford, M.M. Prediction of sorghum biomass based on image based features derived from time series of UAV images. In Proceedings of the International Geoscience and Remote Sensing Symposium (IGARSS), Fort Worth, TX, USA, 23–28 July 2017; pp. 6154–6157.

18. Lewis, B.; Smith, I.; Fowler, M.; Licato, J. The robot mafia: A test environment for deceptive robots. In Proceedings of the 28th Modern Artificial Intelligence and Cognitive Science Conference, MAICS 2017, Fort Wayne, IN, USA, 28–29 April 2017; pp. 189–190.
19. Masjedi, A.; Zhao, J.; Thompson, A.M.; Yang, K.W.; Flatt, J.E.; Crawford, M.M.; Ebert, D.S.; Tuinstra, M.R.; Hammer, G.; Chapman, S. Sorghum biomass prediction using uav-based remote sensing data and crop model simulation. In Proceedings of the International Geoscience and Remote Sensing Symposium (IGARSS), Valencia, Spain, 23–27 July 2018; pp. 7719–7722.
20. Ostos-Garrido, F.J.; de Castro, A.I.; Torres-Sánchez, J.; Pistón, F.; Peña, J.M. High-Throughput Phenotyping of Bioethanol Potential in Cereals Using UAV-Based Multi-Spectral Imagery. *Front. Plant Sci.* **2019**, *10*, 1–15. [CrossRef]
21. Sagan, V.; Maimaitijiang, M.; Sidike, P.; Eblimit, K.; Peterson, K.T.; Hartling, S.; Esposito, F.; Khanal, K.; Newcomb, M.; Pauli, D.; et al. UAV-based high resolution thermal imaging for vegetation monitoring, and plant phenotyping using ICI 8640 P, FLIR Vue Pro R 640, and thermomap cameras. *Remote Sens.* **2019**, *11*, 330. [CrossRef]
22. Holman, F.H.; Riche, A.B.; Castle, M.; Wooster, M.J.; Hawkesford, M.J. Radiometric calibration of "commercial offthe shelf" cameras for UAV-based high-resolution temporal crop phenotyping of reflectance and NDVI. *Remote Sens.* **2019**, *11*, 1657. [CrossRef]
23. Enciso, J.; Avila, C.A.; Jung, J.; Elsayed-Farag, S.; Chang, A.; Yeom, J.; Landivar, J.; Maeda, M.; Chavez, J.C. Validation of agronomic UAV and field measurements for tomato varieties. *Comput. Electron. Agric.* **2019**, *158*, 278–283. [CrossRef]
24. Ampatzidis, Y.; Partel, V. UAV-based high throughput phenotyping in citrus utilizing multispectral imaging and artificial intelligence. *Remote Sens.* **2019**, *11*, 410. [CrossRef]
25. Fernandes, S.B.; Dias, K.O.G.; Ferreira, D.F.; Brown, P.J. Efficiency of multi-trait, indirect, and trait-assisted genomic selection for improvement of biomass sorghum. *Theor. Appl. Genet.* **2018**, *131*, 747–755. [CrossRef] [PubMed]
26. Ogbaga, C.C.; Bajhaiya, A.K.; Gupta, S.K. Improvements in biomass production: Learning lessons from the bioenergy plants maize and sorghum. *J. Environ. Biol.* **2019**, *40*, 400–406. [CrossRef]
27. Prabhakara, K.; Dean Hively, W.; McCarty, G.W. Evaluating the relationship between biomass, percent groundcover and remote sensing indices across six winter cover crop fields in Maryland, United States. *Int. J. Appl. Earth Obs. Geoinf.* **2015**, *39*, 88–102. [CrossRef]
28. Moghimi, A.; Yang, C.; Anderson, J.A. Aerial hyperspectral imagery and deep neural networks for high-throughput yield phenotyping in wheat. *Comput. Electron. Agric.* **2020**, *172*, 105299. [CrossRef]
29. Zhao, J.; Karimzadeh, M.; Masjedi, A.; Wang, T.; Zhang, X.; Crawford, M.M.; Ebert, D.S. FeatureExplorer: Interactive Feature Selection and Exploration of Regression Models for Hyperspectral Images. In Proceedings of the 2019 IEEE Visualization Conference VIS, Vancouver, BC, Canada, 20–25 October 2019; pp. 161–165. [CrossRef]
30. Feng, W.; Guo, B.B.; Zhang, H.Y.; He, L.; Zhang, Y.S.; Wang, Y.H.; Zhu, Y.J.; Guo, T.C. Remote estimation of above ground nitrogen uptake during vegetative growth in winter wheat using hyperspectral red-edge ratio data. *Field Crop. Res.* **2015**, *180*, 197–206. [CrossRef]
31. Foster, A.J.; Kakani, V.G.; Mosali, J. Estimation of bioenergy crop yield and N status by hyperspectral canopy reflectance and partial least square regression. *Precis. Agric.* **2017**, *18*, 192–209. [CrossRef]
32. Yue, J.; Feng, H.; Yang, G.; Li, Z. A comparison of regression techniques for estimation of above-ground winter wheat biomass using near-surface spectroscopy. *Remote Sens.* **2018**, *10*, 66. [CrossRef]
33. Fassnacht, F.E.; Hartig, F.; Latifi, H.; Berger, C.; Hernández, J.; Corvalán, P.; Koch, B. Importance of sample size, data type and prediction method for remote sensing-based estimations of aboveground forest biomass. *Remote Sens. Environ.* **2014**, *154*, 102–114. [CrossRef]
34. Vaglio Laurin, G.; Puletti, N.; Chen, Q.; Corona, P.; Papale, D.; Valentini, R. Above ground biomass and tree species richness estimation with airborne lidar in tropical Ghana forests. *Int. J. Appl. Earth Obs. Geoinf.* **2016**, *52*, 371–379. [CrossRef]
35. Harkel, J.T.; Bartholomeus, H.; Kooistra, L. Biomass and crop height estimation of different crops using UAV-based LiDAR. *Remote Sens.* **2020**, *12*, 17. [CrossRef]

36. McGlinchy, J.; Van Aardt, J.A.N.; Erasmus, B.; Asner, G.P.; Mathieu, R.; Wessels, K.; Knapp, D.; Kennedy-Bowdoin, T.; Rhody, H.; Kerekes, J.P.; et al. Extracting structural vegetation components from small-footprint waveform lidar for biomass estimation in savanna ecosystems. *IEEE J. Sel. Top. Appl. Earth Obs. Remote Sens.* **2014**, *7*, 480–490. [CrossRef]
37. Shao, G.; Shao, G.; Gallion, J.; Saunders, M.R.; Frankenberger, J.R.; Fei, S. Improving Lidar-based aboveground biomass estimation of temperate hardwood forests with varying site productivity. *Remote Sens. Environ.* **2018**, *204*, 872–882. [CrossRef]
38. Phua, M.H.; Johari, S.A.; Wong, O.C.; Ioki, K.; Mahali, M.; Nilus, R.; Coomes, D.A.; Maycock, C.R.; Hashim, M. Synergistic use of Landsat 8 OLI image and airborne LiDAR data for above-ground biomass estimation in tropical lowland rainforests. *For. Ecol. Manag.* **2017**, *406*, 163–171. [CrossRef]
39. Vastaranta, M.; Holopainen, M.; Karjalainen, M.; Kankare, V.; Hyyppa, J.; Kaasalainen, S. TerraSAR-X stereo radargrammetry and airborne scanning LiDAR height metrics in imputation of forest aboveground biomass and stem volume. *IEEE Trans. Geosci. Remote Sens.* **2014**, *52*, 1197–1204. [CrossRef]
40. Zhao, K.; Suarez, J.C.; Garcia, M.; Hu, T.; Wang, C.; Londo, A. Utility of multitemporal lidar for forest and carbon monitoring: Tree growth, biomass dynamics, and carbon flux. *Remote Sens. Environ.* **2018**, *204*, 883–897. [CrossRef]
41. Zhu, Y.; Zhao, C.; Yang, H.; Yang, G.; Han, L.; Li, Z.; Feng, H.; Xu, B.; Wu, J.; Lei, L. Estimation of maize above-ground biomass based on stem-leaf separation strategy integrated with LiDAR and optical remote sensing data. *PeerJ* **2019**, *7*, 1–30. [CrossRef]
42. Luo, S.; Wang, C.; Xi, X.; Nie, S.; Fan, X.; Chen, H.; Yang, X.; Peng, D.; Lin, Y.; Zhou, G. Combining hyperspectral imagery and LiDAR pseudo-waveform for predicting crop LAI, canopy height and above-ground biomass. *Ecol. Indic.* **2019**, *102*, 801–812. [CrossRef]
43. Chao, Z.; Liu, N.; Zhang, P.; Ying, T.; Song, K. Estimation methods developing with remote sensing information for energy crop biomass: A comparative review. *Biomass Bioenergy* **2019**, *122*, 414–425. [CrossRef]
44. Vaglio Laurin, G.; Chen, Q.; Lindsell, J.A.; Coomes, D.A.; Frate, F.D.; Guerriero, L.; Pirotti, F.; Valentini, R. Above ground biomass estimation in an African tropical forest with lidar and hyperspectral data. *ISPRS J. Photogramm. Remote Sens.* **2014**, *89*, 49–58. [CrossRef]
45. Ravi, R.; Lin, Y.J.; Elbahnasawy, M.; Shamseldin, T.; Habib, A. Simultaneous System Calibration of a Multi-LiDAR Multicamera Mobile Mapping Platform. *IEEE J. Sel. Top. Appl. Earth Obs. Remote Sens.* **2018**, *11*, 1694–1714. [CrossRef]
46. LaForest, L.; Hasheminasab, S.M.; Zhou, T.; Flatt, J.E.; Habib, A. New strategies for time delay estimation during system calibration for UAV-Based GNSS/INS-Assisted imaging systems. *Remote Sens.* **2019**, *11*, 1811. [CrossRef]
47. He, F.; Zhou, T.; Xiong, W.; Hasheminnasab, S.M.; Habib, A. Automated aerial triangulation for UAV-based mapping. *Remote Sens.* **2018**, *10*, 1952. [CrossRef]
48. Hasheminasab, S.M.; Zhou, T.; Habib, A. GNSS/INS-Assisted structure from motion strategies for UAV-Based imagery over mechanized agricultural fields. *Remote Sens.* **2020**, *12*, 351. [CrossRef]
49. Habib, A.; Zhou, T.; Masjedi, A.; Zhang, Z.; Evan Flatt, J.; Crawford, M. Boresight Calibration of GNSS/INS-Assisted Push-Broom Hyperspectral Scanners on UAV Platforms. *IEEE J. Sel. Top. Appl. Earth Obs. Remote Sens.* **2018**, *11*, 1734–1749. [CrossRef]
50. Liu, Y.-K.; Li, C.-R.; Ma, L.-L.; Qian, Y.-G.; Wang, N.; Gao, C.-X.; Tang, L.-L. Land surface reflectance retrieval from optical hyperspectral data collected with an unmanned aerial vehicle platform. *Opt. Express* **2019**, *27*, 7174. [CrossRef]
51. Thorp, K.R.; Wang, G.; Bronson, K.F.; Badaruddin, M.; Mon, J. Hyperspectral data mining to identify relevant canopy spectral features for estimating durum wheat growth, nitrogen status, and grain yield. *Comput. Electron. Agric.* **2017**, *136*, 1–12. [CrossRef]
52. Demetriades-Shah, T.H.; Steven, M.D.; Clark, J.A. High resolution derivative spectra in remote sensing. *Remote Sens. Environ.* **1990**, *33*, 55–64. [CrossRef]
53. Feng, W.; Guo, B.B.; Wang, Z.J.; He, L.; Song, X.; Wang, Y.H.; Guo, T.C. Measuring leaf nitrogen concentration in winter wheat using double-peak spectral reflection remote sensing data. *Field Crop. Res.* **2014**, *159*, 43–52. [CrossRef]
54. Savitzky, A.; Golay, M.J.E. Smoothing and Differentiation of Data by Simplified Least Squares Procedures. *Anal. Chem.* **1964**, *36*, 1627–1639. [CrossRef]

55. Asner, G.P.; Martin, R.E. Spectral and chemical analysis of tropical forests: Scaling from leaf to canopy levels. *Remote Sens. Environ.* **2008**, *112*, 3958–3970. [CrossRef]
56. Zhao, Y.R.; Li, X.; Yu, K.Q.; Cheng, F.; He, Y. Hyperspectral Imaging for Determining Pigment Contents in Cucumber Leaves in Response to Angular Leaf Spot Disease. *Sci. Rep.* **2016**, *6*, 1–9. [CrossRef]
57. Féret, J.B.; François, C.; Gitelson, A.; Asner, G.P.; Barry, K.M.; Panigada, C.; Richardson, A.D.; Jacquemoud, S. Optimizing spectral indices and chemometric analysis of leaf chemical properties using radiative transfer modeling. *Remote Sens. Environ.* **2011**, *115*, 2742–2750. [CrossRef]
58. Ullah, S.; Skidmore, A.K.; Ramoelo, A.; Groen, T.A.; Naeem, M.; Ali, A. Retrieval of leaf water content spanning the visible to thermal infrared spectra. *ISPRS J. Photogramm. Remote Sens.* **2014**, *93*, 56–64. [CrossRef]
59. Thulin, S.; Hill, M.J.; Held, A.; Jones, S.; Woodgate, P. Predicting Levels of Crude Protein, Digestibility, Lignin and Cellulose in Temperate Pastures Using Hyperspectral Image Data. *Am. J. Plant Sci.* **2014**, *05*, 997–1019. [CrossRef]
60. Ecarnot, M.; Compan, F.; Roumet, P. Assessing leaf nitrogen content and leaf mass per unit area of wheat in the field throughout plant cycle with a portable spectrometer. *Field Crop. Res.* **2013**, *140*, 44–50. [CrossRef]
61. Li, X.; Zhang, Y.; Bao, Y.; Luo, J.; Jin, X.; Xu, X.; Song, X.; Yang, G. Exploring the best hyperspectral features for LAI estimation using partial least squares regression. *Remote Sens.* **2014**, *6*, 6221–6241. [CrossRef]
62. Zhang, T. *An Introduction to Support Vector Machines and Other Kernel-Based Learning Methods A Review*; Cambridge University Press: Cambridge, UK, 2001; Volume 22, ISBN 0521780195.
63. Breiman, L. Random forests. *Mach. Learn.* **2001**, *45*, 5–32. [CrossRef]
64. Belgiu, M.; Drăgu, L. Random forest in remote sensing: A review of applications and future directions. *ISPRS J. Photogramm. Remote Sens.* **2016**, *114*, 24–31. [CrossRef]
65. Blondel, M.; Brucher, M.; Buitinck, L.; Cournapeau, D.; Dawe, N.; Du, S.; Dubourg, V.; Duchesnay, E.; Fabisch, A.; Fritsch, V.; et al. Scikit-learn. *J. Mach. Learn. Res.* **2015**, *12*, 2825–2830. [CrossRef]
66. Sokal, R.R.; James Rohlf, F. *Biometry: The Principles and Practice of Statistics in Biological Research*; W. H. Freeman: New York, NY, USA, 1995.
67. Seabold, S.; Perktold, J. Statsmodels: Econometric and Statistical Modeling with Python. In Proceedings of the 9th Python in Science Conference, Austin, TX, USA, 28 June–3 July 2010; pp. 92–96.
68. Gerik, T.; Bean, B.; Vanderlip, R. *Sorghum Growth and Development*; Texas FARMER Collection, Texas Agrilife Extension, Texas A&M University: College Station, TX, USA, 2003.
69. De Almeida, C.T.; Galvão, L.S.; de Aragão, L.E.O.C.e.; Ometto, J.P.H.B.; Jacon, A.D.; de Pereira, F.R.S.; Sato, L.Y.; Lopes, A.P.; de Graça, P.M.L.A.; de Silva, C.V.J.; et al. Combining LiDAR and hyperspectral data for aboveground biomass modeling in the Brazilian Amazon using different regression algorithms. *Remote Sens. Environ.* **2019**, *232*, 111323. [CrossRef]
70. Gitelson, A.; Merzlyak, M.N. Quantitative estimation of chlorophyll-a using reflectance spectra: Experiments with autumn chestnut and maple leaves. *J. Photochem. Photobiol. B Biol.* **1994**, *22*, 247–252. [CrossRef]
71. Marshak, A.; Knyazikhin, Y.; Davis, A.B.; Wiscombe, W.J.; Pilewskie, P. Cloud-vegetation interaction: Use of normalized difference cloud index for estimation of cloud optical thickness. *Geophys. Res. Lett.* **2000**, *27*, 1695–1698. [CrossRef]
72. Carter, G.A. Ratios of leaf reflectances in narrow wavebands as indicators of plant stress. *Int. J. Remote Sens.* **1994**, *15*, 517–520. [CrossRef]
73. Sims, D.A.; Gamon, J.A. Relationships between leaf pigment content and spectral reflectance acrossa wide range of species, leaf structures and developmental stages. *Int. J. Remote Sens.* **2002**, *81*, 337–354. [CrossRef]
74. Gitelson, A.A.; Kaufman, Y.J.; Merzlyak, M.N. Use of a green channel in remote sensing of global vegetation from EOS-MODIS. *Remote Sens. Environ.* **1996**, *58*, 289–298. [CrossRef]
75. McMurtrey, J.E.; Chappelle, E.W.; Kim, M.S.; Meisinger, J.J.; Corp, L.A. Distinguishing nitrogen fertilization levels in field corn (Zea mays L.) with actively induced fluorescence and passive reflectance measurements. *Remote Sens. Environ.* **1994**, *47*, 36–44. [CrossRef]
76. Rondeaux, G.; Steven, M.; Baret, F. Optimization of soil-adjusted vegetation indices. *Remote Sens. Environ.* **1996**, *55*, 95–107. [CrossRef]
77. Daughtry, C.S.T.; Walthall, C.L.; Kim, M.S.; de Colstoun, E.B.; McMurtrey, J.E. Estimating Corn Leaf Chlorophyll Concentration from Leaf and Canopy Reflectance. *Remote Sens. Environ.* **2000**, *35*, 229–239. [CrossRef]

78. Clevers, J.G.P.W. *Imaging Spectrometry in Agriculture—Plant Vitality And Yield Indicators BT—Imaging Spectrometry—A Tool for Environmental Observations*; Hill, J., Mégier, J., Eds.; Springer: Dordrecht, The Netherlands, 1994; pp. 193–219. ISBN 978-0-585-33173-7.
79. Gamon, J.A.; Peñuelas, J.; Field, C.B. A Narrow-Waveband Spectral Index That Tracks Diurnal Changes in Photosynthetic Efficiency. *Remote Sens. Environ.* **1992**, *44*, 35–44. [CrossRef]
80. Gao, B.-C. NDWI A Normalized Difference Water Index for Remote Sensing of Vegetation Liquid Water From Space. *Remote Sens. Environ.* **1996**, *58*, 257–266. [CrossRef]
81. Serrano, L.; Peñuelas, J.; Ustin, S.L. Remote sensing of nitrogen and lignin in Mediterranean vegetation from AVIRIS data: Decomposing biochemical from structural signals. *Remote Sens. Environ.* **2002**, *81*, 355–364. [CrossRef]

MDPI
St. Alban-Anlage 66
4052 Basel
Switzerland
Tel. +41 61 683 77 34
Fax +41 61 302 89 18
www.mdpi.com

Remote Sensing Editorial Office
E-mail: remotesensing@mdpi.com
www.mdpi.com/journal/remotesensing

www.ingramcontent.com/pod-product-compliance
Lightning Source LLC
LaVergne TN
LVHW070903170726
843515LV00005B/695

* 9 7 8 3 0 3 6 5 0 5 2 6 8 *